Springer Natural Hazards

The Springer Natural Hazards series seeks to publish a broad portfolio of scientific books, aiming at researchers, students, and everyone interested in Natural Hazard research. The series includes peer-reviewed monographs, edited volumes, textbooks, and conference proceedings. It covers all categories of hazards such as atmospheric/climatological/oceanographic hazards, storms, tsunamis, floods, avalanches, landslides, erosion, earthquakes, volcanoes, and welcomes book proposals on topics like risk assessment, risk management, and mitigation of hazards, and related subjects.

Yujing Jiang • Xun Li

Stability Evaluation of Slope Under Heavy Rainfall Based on Geometric Features

Springer

Yujing Jiang
Graduate School of Engineering
Nagasaki University
Nagasaki, Nagasaki, Japan

Xun Li
School of Water Resources and Hydropower Engineering
Wuhan University
Wuhan, China

ISSN 2365-0656 ISSN 2365-0664 (electronic)
Springer Natural Hazards
ISBN 978-981-95-8655-4 ISBN 978-981-95-8656-1 (eBook)
https://doi.org/10.1007/978-981-95-8656-1

This Springer imprint is published by the registered company Springer Nature Singapore Pte Ltd.
The registered company address is: 152 Beach Road, #21-01/04 Gateway East, Singapore 189721, Singapore

Preface

In rainfall-induced slope stability analyses, discrepancies often exist between numerical simulation results and the actual behavior of natural slopes. These discrepancies are largely attributable to the common assumption of uniform physical and mechanical properties, which neglects the inherent heterogeneity of slope materials. To address this limitation, this study introduces a statistical approach to characterize the heterogeneity of weathered layers, in which the mechanical properties are described using a Weibull distribution. Based on this framework, coupled unsaturated–saturated seepage and slope stability analyses under heavy rainfall were conducted. To enhance simulation accuracy, the parameters of the van Genuchten (VG) model and the fluid modulus K_f were determined through laboratory tests. A case study demonstrates that the Weibull-based representation effectively captures varying degrees of heterogeneity in weathered layers, yielding more conservative and realistic stability assessments under heavy rainfall. In addition to material heterogeneity, rainfall-induced weakening of soft rocks is another critical factor influencing slope stability. To quantify this effect, FISH functions in FLAC^{3D} were employed to investigate the influence of porosity and heterogeneity on the peak strength of sandstone under uniaxial compression. This approach was further extended to evaluate stability variations in a two-dimensional (2D) slope subjected to heavy rainfall. A comparative analysis was conducted between two modeling strategies: (i) simplifying mechanical properties into dry and fully saturated states and (ii) dynamically updating material parameters according to real-time saturation levels. The results underscore the necessity of accounting for both material heterogeneity and the dynamic evolution of rock properties during rainfall infiltration.

The study further examines the influence of surface geometry on the stability of three-dimensional (3D) slopes and assesses the limitations of using 2D cross-sections to represent 3D slope behavior. By systematically varying the amplitude and frequency of surface undulations as well as slope gradients, the analyses reveal that 3D slopes with pronounced, high-frequency surface roughness generally exhibit lower factors of safety (FoS) than their 2D counterparts. These findings highlight the significant role of spatial dimensionality in governing the stability of large and geometrically complex slopes, indicating that a combined 2D–3D analytical framework is essential for reliable assessment. Finally, an efficient method is proposed for rapid evaluation of the FoS of 3D slopes under heavy rainfall by linking geometric characteristics to slope stability. Given the high computational cost of fully coupled 3D unsaturated seepage–stability analyses, reliance on representative 2D sections may lead to substantial inaccuracies for complex terrains. To overcome this limitation, the slope surface roughness coefficient (SSRC) is introduced as a quantitative index to characterize the roughness of 3D slope surfaces. By establishing a relationship between SSRC and slope stability, the proposed approach enables fast and reliable estimation of the FoS of 3D slopes. Comparative analyses with detailed 3D numerical models demonstrate that SSRC serves as a robust indicator of slope stability under heavy rainfall. Validation against comprehensive FLAC^{3D} seepage–stability simulations confirms that the predicted FoS values closely match the numerical results. This agreement underscores the practical applicability and reliability of the SSRC-based method.

Nagasaki, Japan Yujing Jiang
Wuhan, China Xun Li

Acknowledgments

The numerical simulation, spatial analysis platform, and experimental facilities are provided by Nagasaki University. The authors appreciate their essential support. The authors gratefully acknowledge the Road Maintenance Division, Nagasaki Prefecture for providing technical references related to disaster sites and cooperating during on-site investigations.

We wish to thank the staff at Springer Nature Press, in particular Yosuke Nishida, Sridevi Purushothaman, and Mahalakshmi Saravanan, for their kind help and assistance.

January 2026

Yujing Jiang
Xun Li

Competing Interests The authors have no competing interests to declare that are relevant to the content of this manuscript.

Contents

Chapter 1
Introduction

Abstract This chapter begins by outlining the research background of rainfall-induced landslides. It focuses on two critical issues in this field, namely the heterogeneity of slope geomaterials and the influence of water-induced weakening effects in soft rock on slope stability during rainfall. The discussion then turns to the limitations of widely adopted GIS-based methods for regional-scale analysis of rainfall-induced slope stability. Drawing on this background, the chapter presents a literature review that synthesizes previous studies and highlights key developments in the analysis of slope stability under rainfall conditions. The review covers three main aspects, including two-dimensional (2D) and three-dimensional (3D) slope stability analysis under rainfall conditions, and regional-scale assessment of rainfall-induced landslide susceptibility.

Keywords Subtropical regions · Rainfall-induced landslides · 2D slope stability · 3D slope stability · Regional-scale assessment

1.1 Background

Rainfall is a critical factor triggering slope failures and nearly 450 rainfall-induced landslides were reported around the world in 2016 (Sheikh et al. 2021). According to the research of Froude and Petley (2018), rainfall-triggered landslides are globally distributed, with the majority occurring in tropical and subtropical regions where steep terrain is prevalent. This trend is supported by data from the Global Fatal Landslide Database (GFLD), as shown in Fig. 1.1. Asia accounted for 75% of all rainfall-induced landslides between 2004 and 2018 (Amarasinghe et al. 2024). In Japan, short-term heavy rainfall events are more frequent in the summer, and the average annual number of rainfall events with an intensity greater than 50 mm/h has been steadily increasing over the past decades (Chueasamat et al. 2018). Their recorded data also indicates that the number of landslides occurring in the summer is significantly higher than in other seasons. The landslides caused by heavy

Y. Jiang, X. Li, *Stability Evaluation of Slope Under Heavy Rainfall Based on Geometric Features*, Springer Natural Hazards,
https://doi.org/10.1007/978-981-95-8656-1_1

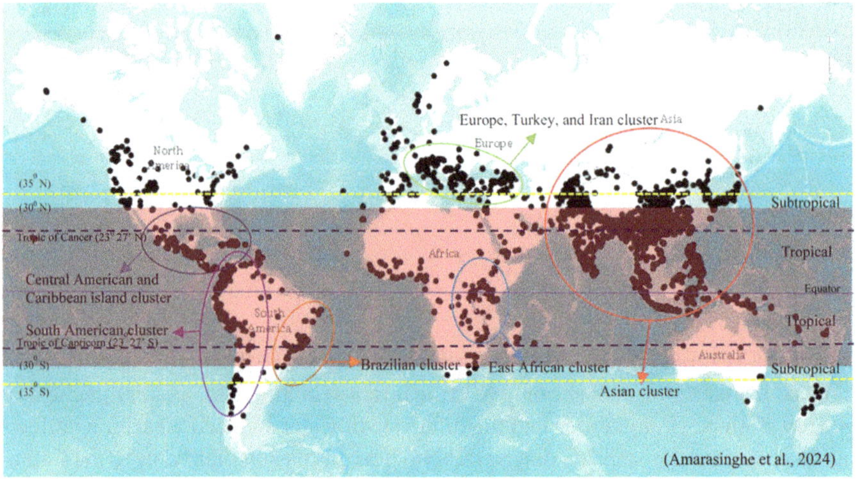

Fig. 1.1 Rain-induced landslides distribution from 2004 to 2018 based on the GFLD

Fig. 1.2 A rainfall-induced landslide occurred in Akakubi Town, Nagasaki City, Japan in 2021

rainstorms in June–July 2018 and July 2020 resulted in significant loss of lives and properties in Japan (Kitamura and Sako 2010; Cho et al. 2021). As shown in Fig. 1.2, a landslide occurred in Akakubi Town, Nagasaki City, Japan, in August 2021 due to heavy rainfall. From August 10 to August 17, the cumulative rainfall amount is 939 mm, the maximum 24-h rainfall amount is 286 mm, and the maximum hourly rainfall is 50 mm. This continuous heavy rainfall caused landslides and threatened

Fig. 1.3 Landslides induced by heavy rainfall on November 6, 2024 in Florida, USA

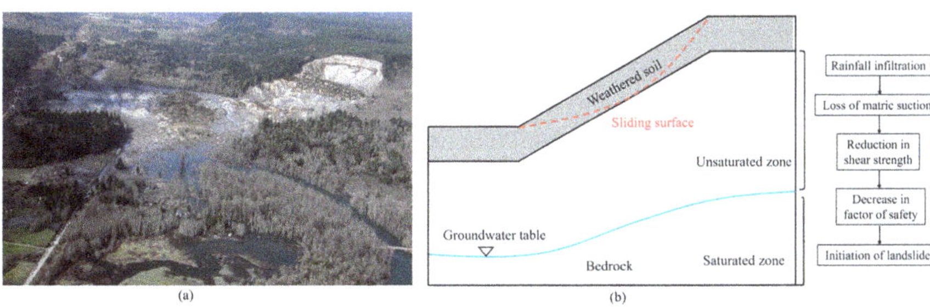

Fig. 1.4 (**a**) A rainfall-induced landslide in northwest Washington on 2014. (**b**) The mechanism of rainfall-induced landslide

people's lives. Recently, the hurricane traversed the southern Appalachian Mountains, bringing torrential rainfall that triggered widespread landslides across Florida, USA. Notably, on November 6, 2024, heavy rainfall in Florida caused multiple rainfall-induced landslides, resulting in significant impacts on the affected areas, as illustrated in Fig. 1.3. Figure 1.4a shows a typical rainfall-induced landslide occurred on 2014, in northwest Washington, USA. It illustrates the entire source zone and runout trajectory of the landslide. Therefore, investigating the mechanisms of triggering rainfall-induced landslides holds significant importance for the prevention of geological hazards caused by slope failures. The fundamental mechanism of shallow rainfall-induced landslides is illustrated in Fig. 1.4b. Rainfall infiltration reduces matric suction in the unsaturated zone of a slope (Cai and Ugai

2004). The increase in soil moisture also raises the self-weight of soils and decreases its effective stress. As a result, the shear strength is reduced, which lowers the overall slope stability (Ng and Shi 1998). Numerical simulations, such as the finite difference method (FDM) (Xu et al. 2017; Pan et al. 2020) and the finite element method (FEM) (Oh and Lu 2015; Rafiei Renani and Martin 2020), are commonly employed to analyze slope stability under rainfall conditions. In most of these studies, the geomechanical parameters of slope materials are assumed to be uniform. This approach fails to capture the inherent heterogeneity of actual slopes (Cho 2014). Because weathered layers often exhibit weaker mechanical properties than the average, simulations using uniform parameters may predict a safety factor greater than 1 even when slope failure occurs in reality (Chen et al. 2021; Li et al. 2025).

Water content is an important factor in reducing rock strength, and even a slight increase in water content can significantly decrease the strength and stiffness of sedimentary rocks (Yilmaz 2010). Numerous studies have demonstrated that the stability of rock slopes can be significantly affected by the weakening effects caused by the interaction between water and rock (Vásárhelyi and Ván 2006; Li et al. 2012; Iyare et al. 2023). These studies have primarily focused on the influence of water on the uniaxial compressive strength (UCS) of various sedimentary rocks, including sandstone, mudstone, and siltstone. A negative exponential relationship is typically observed between UCS and water content across these rock types, showing that as water content increases, UCS decreases exponentially and highlighting the weakening effect of water on their mechanical strength (Vásárhelyi and Ván 2006). The results reveal that both the peak strength and elastic modulus of red sandstone decrease with increasing water content and immersion time, and these changes can be described by a negative exponential function (Yu et al. 2019). In the past decades, most research has focused on understanding how water reduces the mechanical strength of various rock types (Duperret et al. 2005; Erguler and Ulusay 2009; Wasantha and Ranjith 2014). However, few studies have examined the water-induced softening of heterogeneous rocks. The analysis of slope stability under rainfall conditions has been a challenging research topic in geotechnical engineering over the past few decades (Xiong et al. 2019). One of the primary challenges lies in simulating the heterogeneity of slope materials in numerical simulations.

Furthermore, most existing GIS-based methods primarily utilize either a one-dimensional (1D) infinite slope model (Gutiérrez-Martín 2020) to analyze rainfall infiltration or a two-dimensional (2D) slope model derived from cross-sections of a digital elevation model (DEM) for stability analysis (Gu et al. 2015). Although 1D and 2D approaches are often preferred for their simplicity and perceived conservative estimates, they tend to underestimate slope stability when compared to 3D methods (Tran et al. 2017). This underestimation occurs because 1D and 2D analyses typically simplify complex geological and hydrological conditions, leading to less accurate risk assessments for landslides. When specific combinations of soil and groundwater characteristics, complex geometries (e.g., ridges and corners), and boundary conditions are considered in the numerical simulation, it is highly recommended to conduct the 3D slope stability analysis instead of only calculating the 2D critical cross-sections (Li et al. 2024). However, three-dimensional (3D) slope stability assessment at a regional level is not commonly practiced due to its complexity, high computational demands, and lack of sufficient data (Tran et al. 2017). Therefore, it is essential to propose a novel, more accurate and efficient method for regional 3D slope stability analysis based on a detailed analysis of the impact of unsaturated seepage on slope stability under rainfall infiltration conditions.

1.2 Literature Review on Slope Stability Analysis Under Rainfall Conditions

1.2.1 2D Slope Stability Analysis Under Rainfall Conditions

In the past decades, the most common approach for slope stability analysis under rainfall infiltration is the application of numerical simulation such as the finite difference method (FDM) (Xu et al. 2017; Pan et al. 2020) and the finite element method (FEM) (Cai and Ugai 2004; Cho and Lee 2001; Griffiths and Lane 1999; Ng and Shi 1998; Oh and Lu 2015; Rafiei Renani and Martin 2020). A finite element analysis was performed to investigate the transient seepage of saturated-unsaturated soil under various hydraulic characteristics and the effects of rainfall intensity and duration on the slope stability (Cai and Ugai 2004; Ng and Shi 1998). However, the transient seepage analyses (assuming non-deforming soil) and slope stability calculations were conducted in a completely uncoupled manner, and potentially leading to the inaccuracy in the calculation with neglecting fluid-solid interaction. In fact, the deformation and stability changes in slopes during the rainfall infiltration process are closely related to the coupling effects of solid and liquid (Xu et al. 2021; Wang et al. 2023). The alteration of the soil's microstructure due to deformation influences the pathways of water migration, permeability characteristics, and the overall seepage process. It is imperative to consider the impact of fluid-solid coupling effects in unsaturated seepage analyses to enhance the accuracy of the calculation. Although the finite element software was employed to conduct fluid-solid coupling analysis on a clastic rock slope under rainfall infiltration (Xu et al. 2017), limitations existed in their research without considering the real-time changes in the permeability and shear strength of the rock-soil mass under rainfall conditions. These issues can be effectively addressed by utilizing FISH functions for secondary development within the $FLAC^{3D}$. For example, Pan et al. (2020) proposed a feasible and more accurate method for analyzing the stability of rock slopes under rainfall infiltration based on the secondary development of FISH functions. They focused on changes of pore pressure in the slope due to rainfall and the reduction in shear strength caused by water-induced weakening effects of rocks. Through secondary development using FISH functions, they effectively incorporated real-time corrections to the permeability of the rock mass during the unsaturated seepage process. However, the impact of matric suction on shear strength in the weathered layer of the rock-soil mass was not considered. In contrast to soil slopes, they have illustrated that the impact of water on the instability of rock slopes not only attributes to changes in density and pore pressure. The presence of water significantly diminishes the strength and stiffness of the rock. Consequently, water-induced weakening effects play a more substantial role in influencing rock slopes compared to their impact on soil slopes.

The assessment of slope stability is crucial in geotechnical engineering, especially in areas prone to slope failures caused by rainfall infiltration. This research outlined a systematic procedure for determining the safety factor of unsaturated slopes affected by rainfall infiltration (Cho and Lee 2001). They employed a two-dimensional finite element analysis program to explore the interaction between rainfall infiltration and the behavior of soil slopes. To account for the impact of matric suction on the shear strength of unsaturated soil, a modified version of the Mohr–Coulomb failure criterion was integrated into the analysis. Although they developed flow-deformation coupled analysis program, the real-time variation of permeability coefficient in the soil slope was not considered in their research. In comparison to the calculation method of safety factor they employed, the shear strength reduction method (SSRM) is a more convenient approach as it eliminates the need for a predefined failure plane of the slope. The strength reduction technique has progressively emerged as a practical substitute for the limit equilibrium method since its initial application by Zienkiewicz et al. (1975). This method commonly employs the Mohr-Coulomb failure criterion and involves systematically reducing the shear strength of the material until the slope reaches a state of near equilibrium (Sun et al. 2017). One of the clear advantages of the strength reduction technique is that it does not require a pre-defined failure plane of the slope, failure occurs in the region where the shear stress in the rock-soil mass exceeds the shear strength (Zhao et al. 2015; Tschuchnigg et al. 2015). Accurately simulating the heterogeneity of slope materials in numerical models is a key challenge (Cascini et al. 2010; Li et al. 2025). Using uniform material properties often leads to inaccurate results, as actual slopes contain weathered layers with variable properties (Xu et al. 2008; Hicks et al. 2014). Weathered soils are heterogeneous, composed of minerals, grains, cracks, and voids with distinct properties due to weathering (Ietto et al. 2018). The Weibull distribution is widely used to model heterogeneous rock strength (Tang et al. 2016; Tang and Tang 2012; Vanmarcke and Grigoriu 1983). In numerical simulations, this is often achieved by assigning different mechanical properties to microstructural elements (Tang et al. 2012, 2016; Fang and Harrison 2002). At the macroscale, normal and exponential distributions are commonly used to characterize slope material variability (Cho 2010; Ching and Wang 2016; Jiang et al. 2022).

Numerous studies have demonstrated that the stability of rock slopes can be significantly affected by the weakening effects caused by the interaction between water and rock (Vásárhelyi and Ván 2006; Li et al. 2012; Iyare et al. 2023). In rock mechanics and engineering projects, rocks are commonly classified based on their dry uniaxial compressive strength (UCS). However, for precise engineering design, it is essential to assess both the wet strength and water sensitivity of rocks to accurately account for potential variations in strength and deformability (Li et al. 2012).

In the past decade, considerable research has focused on understanding how water weakens the mechanical properties of various types of rock (Duperret et al. 2005; Erguler and Ulusay 2009; Wasantha and Ranjith 2014). It has been demonstrated that the physico-chemical interactions between rock particles and water are significantly influenced by factors such as mineralogical composition, grain size, pore volume, and other microstructural properties (Wasantha and Ranjith 2014). These characteristics vary widely among different rock types, which plays a crucial role in determining how water affects their mechanical properties. For example, a triaxial compression test was conducted on mudstone specimens to investigate how water content affects their failure behavior (Iyare et al. 2023). The results indicated that the water-weakening effect significantly lowers the depth at which the brittle-ductile transition occurs in mudstones, highlighting the critical role of water in influencing rock deformation and failure behavior (Iyare et al. 2021). Partial or full saturation of mudstone leads to increased deformability, reduced compressive strength, and decreased stiffness (Iverson 2000). This results in increased creep, plastic deformation, and faster failure rates in the triaxial compression test (Brantut et al. 2013). During uniaxial compression testing, the level of damage intensifies as the water content increases, with the damage mode transitioning from single-bevel shear damage and 'X'-type conjugate bevel shear damage to bottom drum-type damage (Jiang et al. 2024). Tang et al. (2000) examined the effects of water-weakening on soft rock by analyzing how the mechanical properties of shale samples change over different immersion periods. Noël et al. (2021) provides new insights into the impact of water on sandstone strength through fracture toughness, frictional, and uniaxial compression tests conducted on five different sandstones under both dry and water-saturated conditions. Their study aims to identify the key physical processes responsible for the short-term and long-term weakening of calcarenites due to water interaction (Ciantia et al. 2015). In general, exposure to water decreases brittleness and increases ductility of soft rocks, resulting in a significant reduction in strength compared to dry conditions, which can contribute to rock engineering hazards such as landslides (Castagna et al. 2018; Iyare et al. 2021). In a saturated state, rock porosity directly affects the water content within the rock, demonstrating a clear exponential relationship between porosity and rock strength (Pan et al. 2020). This result aligns with the conclusions drawn by Li et al. (2012). They also noted that variations in rock porosity result in different water contents at saturation, leading to significant differences in uniaxial compressive strength among samples of the same rock type (Althaus et al. 1994; Kim et al. 2017). The presence of water increases pore pressure within rock and soil masses, which reduces the effective stress on the skeleton particles and alters their physical and mechanical properties (Zhao et al. 2018). Additionally, water can dissolve or modify mineral compositions within these materials, as well as the cement between grains, potentially leading to the formation of new mineral phases (Zhang et al. 2023).

Weathering of rocks is the disintegration and decomposition of minerals, which is a complicated process driven by physical, chemical, and biological interactions (Miščević and Vlastelica 2014). Physical weathering leads to the fragmentation of rocks without altering their mineral composition, while chemical weathering results in the breakdown of the constituent minerals into secondary mineral products (Olona et al. 2010). As these processes alter the original properties and characteristics of rocks, they have shown the significant influence on slope stability analysis (Kim et al. 2015). Miščević and Vlastelica (2014) discussed the impact of weathering on slope stability in soft rock mass. The article covered the various processes associated with weathering, including drying and wetting, and the influence of water pressure on shear strength along rock joints. They concluded that weathering can significantly affect the stability of slopes excavated in soft rocks, and that it is important to consider these factors in stability analyses. Numerous studies categorize weathering horizons into six distinct classes, ranging from the fresh bedrock (Class I) to the fully weathered rock or residual soils (Class VI) (Perri et al. 2016). Slope stability is intrinsically tied to the degree of rock weathering. Weathering-induced changes in rock shear strength and porosity directly influence the stability of slopes. The altered mechanical properties of weathered rock can lead to slope failures (Ersöz and Topal 2018). In recent years, the relationship between rock weathering and slope stability has spurred increased interest among geotechnical engineers. Varying degrees of weathered rocks result in changes of mechanical and hydraulic properties in the slope. Heterogeneity is a critical aspect associated with weathered rock slopes. It refers to the variations in rock properties, such as shear strength, permeability, and deformation characteristics of the rock mass. According to the description by Ietto et al. (2018), for the completely and highly weathered rock masses, the introduction of a fluid-solid coupling model based on the principles of a continuous porous medium is reasonable. The weathered layers of rock slopes cannot be simplified treated to soil or rock. It is the mixture of rock and soil and samples from the study area can typically be fractured using a hammer (Perri et al. 2016; Ietto et al. 2018). They described the rock mass primarily exhibits characteristics resembling soil in a smaller proportion, while a more dominant portion displays properties like rock. Many researchers have explored the feasible approach to simulate this distinct heterogeneity of weathered rocks. A random discrete element method for modeling the spatial variability of rock mechanical properties was performed (Li et al. 2021). A parallel bonding model was utilized to replicate the linking of particles, while adjusting the strength of the bonds in different spatial regions to accommodate the varying composition of the rock. Meng et al. (2019) also proposed a random generation method to simulate the soil and rock mixture (SRM) slope based on the finite element method (FEM) and discrete element method (DEM). They proposed an innovative multiscale strength reduction method employing a hierarchical FEM/DEM framework to evaluate slope stability in the context of heterogeneous granular materials. The Discrete Element Method (DEM) and Bonded Particle Model (BPM) was also employed to simulate the heterogeneity of rock samples (Liakas et al. 2017). The bonded particle model (BPM) was used to replicate heterogeneous rocks composed of distinct structural units characterized by differences in strength and stiffness.

Rocks often exhibit heterogeneity at the mesoscopic scale, resulting in distinct behavior at both the macroscopic and mesoscopic levels due to this heterogeneous nature (Tang et al. 2018; Fang and Harrison 2002). Therefore, it is essential to consider the heterogeneity of rocks in order to improve the accuracy of calculation results (van Mier et al. 2002; Danzer et al. 2007). Weibull (1939) introduced a statistical model, known as the weakest link hypothesis, suggesting that a specimen fails when its weakest volume element fails. This model was proposed to study fractures in brittle materials. From then on, the Weibull distribution has been frequently used to investigate the inhomogeneity characteristics of rocks in the field of rock engineering (Vanmarcke and Grigoriu 1983; Tang and Tang 2012; Tang et al. 2016). Since the heterogeneity of rocks is exhibited at the microscopic level, lots of studies have been conducted on the exploration of failure behavior of rock specimens based on the theory of Weibull distribution (Gulino and Phoenix 1991; Singh and Bhargava 1994; Gupta and Bergström 1998; Tang and Tang 2012; Tang et al. 2016, 2018). The description of heterogeneous characteristics in numerical simulations was achieved by assigning different physical-mechanical parameter values to the microscopic structural units (cells) of rocks (Tang et al. 2018; Tang and Tang 2012). For example, The tensile strength and elastic modulus of rock were assumed to follow the Weibull statistical function to describe the inhomogeneity of rocks in their models (Tang et al. 2012, 2016; Fang and Harrison 2002). The Weibull modulus or shape parameter of the distribution is defined by the homogeneity index m which can be used to describe the degree of rock heterogeneity. The larger value of m indicates a more homogeneous rock (Danzer et al. 2007). However, all these studies concentrated on the failure behavior from the perspective of microstructures of rocks. Few studies were conducted on the application of Weibull distribution from the macroscale of rock slopes. In this study, the mechanical parameters (elastic modulus E, cohesion c) of weathered layers in the rock slope were assigned to follow the Weibull distribution function. In contrast to the emphasis on the research in microscopic levels, the prevailing approach to characterizing the macroscopic statistical patterns of physical-mechanical parameters of rock-soil masses involves normal, exponential distributions and so on (Ching and Wang 2016; Jiang et al. 2022). Suchomel and Mašı'n (2010) considered a spatial random field of uncorrelated parameters c (cohesion) and φ (friction angle) for slope stability analysis. They concluded several limitations of the extended FOSM method for calculating probability of a slope failure, including the assumption of normal distribution of input variables and the neglect of spatial correlation of soil parameters. These investigations were predominantly related to the probabilistic slope stability analysis. Random field theory is extensively employed within stochastic finite element analysis for representing materials exhibiting spatially varying properties in their research (Dou et al. 2014). For instance, Srivastava et al. (2010) conducted a numerical investigation of steady-state seepage flow in the soil slope by regarding the soil permeability as a spatially variable parameter following a correlated log-normal distribution in a two-dimensional domain. The impact of heterogeneous shear strength of soil on the risk and reliability of a long clay slope was investigated (Hicks et al. 2014). They used Monte Carlo simulations and random field theory to model the spatial variability of the clay based on a 3D finite element program.

A numerical procedure for probabilistic slope stability analysis that accounts for uncertainties and spatial variation of soil shear strength parameters was presented

(Kasama and Whittle 2011). The approach was based on the first order reliability method (FORM) and used a two-dimensional Karhunen-Loève expansion to generate random fields of soil properties. The paper discussed the importance of considering spatial variability in slope stability analysis and demonstrated the effectiveness of the proposed method. A random field theory was applied to achieve the heterogeneity of saturated hydraulic conductivity in the rock slope under the rainfall infiltration process (Zhang et al. 2022). By integrating unsaturated seepage theory and random field theory, modified functions for unit saturation, hydraulic conductivity (k), and shear strength parameters were developed for the unsaturated slope. Utilizing FISH and a non-intrusive stochastic method, a saturated-unsaturated seepage random field model was formulated. The model was then employed to analyze the effects of rainfall intensity, duration, and spatial variability of saturated hydraulic conductivity (k_s) on the infiltration process and stability of the unsaturated rock slope. However, the random field is not suitable for the simulation of properties of rocks. Tang et al. (2018) has pointed out that the mechanical properties of rocks should be regarded as following the Weibull distribution. The study highlighted the limitations of the commonly adopted method of simplifying soil-rock slopes as a uniform media and emphasized the importance of considering the distribution and positions of rocks in slope stability analysis. (Liu et al. 2018). They proposed a novel method to analyze the stability of soil-rock slopes that considers the nonuniformity of rocks. Through extracting the original rock position and grading parameters, the corresponding random rock contour lines can be generated based on the finite difference element method. All these studies were mainly focused on the application of random filed theory in the DEM modeling or FEM modeling. However, limited studies have explored the numerical simulation of heterogeneity in shear parameters and the elastic modulus of rock slopes using the finite difference method.

1.2.2 3D Slope Stability Analysis Under Rainfall Conditions

Based on the plane strain assumption, three-dimensional (3D) slope stability can be simplified to a two-dimensional (2D) analysis if one dimension is significantly larger than the others and the slope has no curvature or corners (Zhang et al. 2013; Li et al. 2024). This simplification often yields conservative results, as 2D analyses typically underestimate the factor of safety (FoS). However, field evidence shows that most slope failures are inherently 3D, with complex slip surfaces that violate the plane strain assumption (Jaboyedoff et al. 2020). The assumption of uniform geometry perpendicular to the analysis plane is also inconsistent with actual slopes (Azizian and Popescu 2006; Leong and Rahardjo 2012). Three-dimensional analyses better account for boundary conditions, material heterogeneity, and geological structure orientations (Wines 2016). Consequently, 3D modeling produces results closer to real conditions and more reliable designs, and its feasibility has increased with advances in computational power (Griffiths and Marquez 2007; Wei et al. 2009; Zhang et al. 2015; Zhou et al. 2020). Studies have shown that slope curvature, turning corners, and convex arcs can significantly reduce stability, particularly for steep slopes (Zhang et al. 2013). Moreover, negative pore pressure (matric suction) in unsaturated soils can cause up to 60% differences in FoS between 2D and 3D analyses, with concave geometries showing the largest discrepancies (Zhang et al. 2015). Rainfall infiltration interacts with slope geometry, affecting runoff, initial soil moisture, and subsurface water movement. All of these influence stability (Talebi et al. 2008; Sabzevari and Noroozpour 2014; Wang and Chen 2021). Factors such as slope curvature, plan shape, and width-to-height ratio modify the critical failure surface and FoS, and different rainfall types further impact slope stability (Xu et al. 2021). These findings highlight that 3D analyses are essential for slopes with complex geometry, heterogeneous materials, or unsaturated conditions, while 2D analyses alone may misrepresent actual stability. This study thus focuses on evaluating 3D critical failure surfaces and FoS under rainfall, emphasizing the influence of varying slope surface geometries.

1.2.3 Regional Rainfall-Induced Landslide Susceptibility Assessment

The prediction of regional landslide disasters under heavy rainfall is crucial for effectively preventing geological hazards that arise from slope failures. There are many methods available for predicting regional slope stability under rainfall conditions. Landslide susceptibility assessment methods are primarily categorized into qualitative and quantitative approaches (Park et al. 2013). Qualitative methods rely heavily on expert judgment, making them less suitable for large-scale assessments due to their lack of a solid physical basis in slope failure mechanics. The quantitative methods can be further divided into statistical methods and geotechnical approaches. The geotechnical approach focuses on analyzing the mechanical conditions of slopes and assessing their stability through mathematical calculations, providing a more rigorous framework for evaluating landslide susceptibility (Lee and Chi 2011). In this approach, the physical properties of a particular slope are obtained from field investigating and laboratory tests, and a physical based model is used to evaluate landslide susceptibility. Given the high computational demands associated with regional slope stability analysis, it is widely acknowledged that such complex calculations often require significant computational resources. To streamline the process, making assumptions becomes essential. Different studies prioritize varying factors, carefully selecting assumptions according to their unique goals and analytical frameworks. These assumptions allow for more efficient modeling, even though they may entail some trade-offs in terms of precision or scope (Van Den Bout et al. 2021). They have pointed out that the application of physically-based approaches for slope failure analysis at a catchment scale remains a difficult challenge, and several new models have been proposed in recent years. The assumptions of these models vary significantly. Empirical methods rely on extensive historical data and typically use statistical and regression analyses to derive predictive models. They are suitable for data-rich environments but often struggle to accurately capture extreme events or changes in systems. In contrast, physically-based methods do not rely on historical data but instead simulate actual physical processes to make predictions. This makes them particularly effective in systems where the underlying mechanisms are well understood, especially under extreme conditions. Overall, physically-based methods are crucial in geotechnical engineering problems that demand high precision and dynamic predictions, as they effectively model and simulate real physical processes.

The occurrence of rainfall-induced landslides is influenced by a variety of spatial and climatic factors, including geological composition, topography and hydrogeological conditions. This complexity complicates the prediction of landslide susceptibility, as it requires the collection and analysis of extensive spatial data from the affected region (Park et al. 2013). Consequently, Geographic Information Systems (GIS) have become vital for managing and analyzing spatial information related to landslides, given their robust capabilities in processing distributed data effectively across multiple dimensions (Van Westen et al. 2006). Most existing GIS-based methods typically rely on a one-dimensional (1D) infinite slope model (Gutiérrez-Martín 2020) or a two-dimensional (2D) slope stability model that uses cross-sections of a digital elevation model (DEM) (Gu et al. 2015). Figure 1.5 illustrates a one-dimensional infinite slope model used for GIS-based analysis of regional susceptibility to rainfall-induced landslides. As shown in Fig. 1.5, 'β' is the critical slope angle, 'Z' is the depth of the sliding floor, 'm' is the infiltration factor of water on slope, 'γ_{sat}' stands for the effective weight of saturated soil, 'W' is the weight of the prism considered, '$W.T.$' is the water table and 'P' are opposing shear forces that distort and resist landslides. γ stands for the effective weight of soil, kN/m^3. The calculation of the slope safety factor is provided as follows, where c' represents the effective cohesion and ϕ' represents the effective internal friction angle.

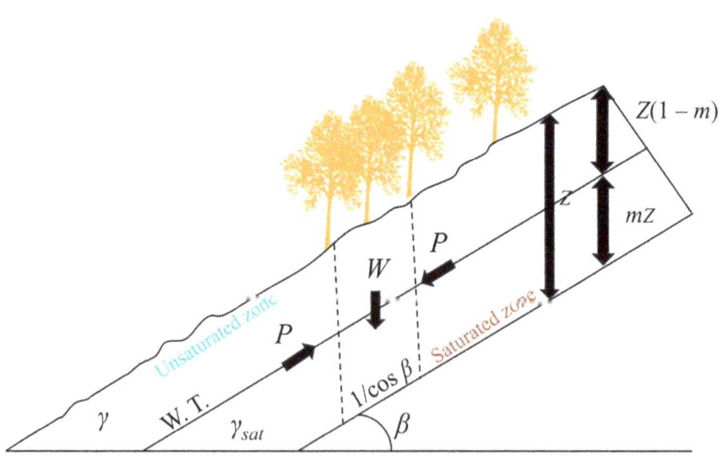

Fig. 1.5 Shallow groundwater conditions in hillside soils; representation of the forces acting on the slope considered by the proposed mathematical infinite-slope model (Gutiérrez-Martín 2020)

$$Fs = \frac{c' + z(\gamma - m\gamma_w)\cos^2\beta \tan\phi'}{\gamma z \sin\beta \cos\beta} \quad (1.1)$$

It is evident that such a simplified one-dimensional seepage model cannot accurately capture the complex processes of unsaturated infiltration of rainfall in 3D slopes. While the one-dimensional model in GIS-based methods simplifies the calculations, it also increases the potential for computational inaccuracies. Although 1D and 2D approaches are often preferred for their simplicity and perceived conservative estimates, they tend to underestimate slope stability when compared to 3D methods (Tran et al. 2017). This underestimation occurs because 1D and 2D analyses typically simplify complex geological and hydrological conditions, leading to less accurate risk assessments for landslides. Since 3D models can provide a more comprehensive evaluation of slope stability and capture the intricate interactions among various factors, such as topography and material properties, relying solely on 1D and 2D methods may result in misleading conclusions about slope stability and associated risks. However, three-dimensional (3D) slope stability assessment at a regional level is not commonly practiced due to its complexity, high computational demands, and lack of sufficient data (Tran et al. 2017). Therefore, most studies on regional rainfall-induced landslide susceptibility, whether employing GIS or other methods, rely on various assumptions to simplify the calculations (Xie et al. 2004; Ho et al. 2012; Park et al. 2013; Jia et al. 2015; Tran et al. 2022).

A column-based 3D limit equilibrium model revised Hovland model was used to calculate the safety factor of each slope unit (Xie et al. 2007). They adopted a Monte-Carlo simulation to locate possible slip surfaces and then applied the revised Hovland model to calculate the corresponding safety factors. Jia et al. (2012) also proposed a method for randomly searching the three-dimensional slope slip surface. To identify the 3D critical slip surface, a Monte Carlo random simulation method is employed to minimize the 3D safety factor. The initial slip surface is assumed to take the form of the lower part of an ellipsoidal slip. Each randomly generated slip surface is subsequently refined to account for variations in stratum strength and the influence of weak discontinuities. This iterative process determines the critical slip surface by achieving a relative minimization of the 3D safety factor. For each slope unit, the critical slip surface is identified using a Monte Carlo simulation, which assumes the slip surface to be the lower part of an ellipsoid (Fig. 1.6a). The simulation employs six parameters to define the size and location of the ellipsoid: the central point (C), the inclination angle (θ), the inclination direction (Asp), and three axial dimensions—semi-minor axis (a), semi-major axis (b), and height (c). The search for the critical slip surface involves iterative trial-and-error adjustments and 3D safety factor calculations. If a generated slip surface based on the lower part of the ellipsoid intersects a weak discontinuity or the boundary of a hard stratum, the weak discontinuity or hard stratum boundary is prioritized as part of the assumed slip surface. As illustrated in Fig. 1.6, the resulting slip surface comprises a portion of the ellipsoid and a portion of the weak discontinuity, reflecting the combined effects of these features. Such assumptions inherently simplify the analysis, as the

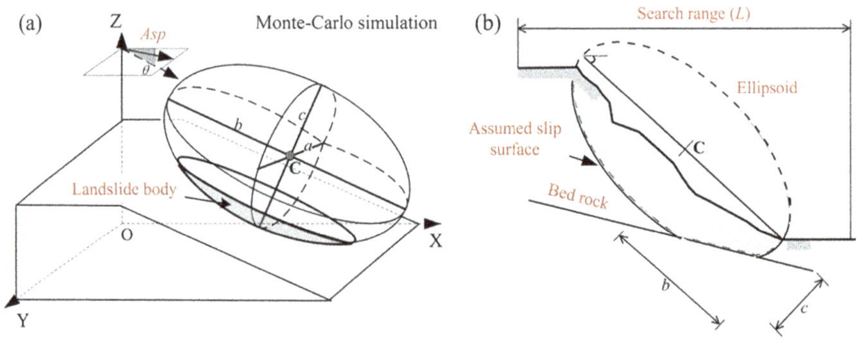

Fig. 1.6 An ellipsoid for slip surface (Jia et al. 2012)

complex slope geometry often results in a failure surface that deviates significantly from an ellipsoidal shape. They have pointed out that a significant drawback of 3D LEM is the challenge of accurately identifying the critical 3D failure surface, both in terms of its shape and location (Wei et al. 2009). Consequently, the random search process across the entire study area, based on the assumption of an ellipsoidal slip surface, is likely to introduce errors in determining the volume and geometry of potential landslides (Van Den Bout et al. 2021). As concluded in their study, the method of simulating an ellipsoid-shaped initial slip surface may overlook certain potential slip surfaces (Xie et al. 2004). Although this random search method for identifying critical slip surfaces is straightforward and easy to implement, it still has several limitations. Firstly, it is computationally expensive and inefficient due to the Monte Carlo simulation's reliance on generating and evaluating numerous slip surfaces. Additionally, the method cannot guarantee finding the global optimum and may converge to a local minimum. The results are often sensitive to initial conditions and simulation parameters, leading to variability and reduced reliability. Its applicability is also limited to complex geological conditions, where the assumed ellipsoidal slip surface may not accurately represent real scenarios. Furthermore, manual intervention is required when the slip surface interacts with weak planes or hard rock boundaries, introducing subjectivity into the process. These limitations collectively hinder the method's efficiency, stability, and adaptability for practical engineering applications.

Both the infinite slope model and revised Hovland model simplify the rainfall infiltration process in unsaturated soils by assuming an idealized scenario, neglecting real-time changes in soil permeability and matric suction during infiltration. These factors are essential for accurately evaluating slope stability under rainfall conditions. Specifically, the model does not consider the soil-water characteristic curve (SWCC), which is crucial for describing the relationship between soil suction and water content. Additionally, the model overlooks the significant effect of matric suction on the shear modulus of soils (Hoyos et al. 2015), which further reduces the accuracy of the factor of safety calculations, potentially leading to less reliable slope stability evaluations. This limitation highlights the need for more sophisticated models that incorporate dynamic changes in soil properties during infiltration, thereby improving the precision of slope stability analyses. The stability analysis of real three-dimensional slopes and the identification of potential sliding zones should also take into account the complex three-dimensional geometric features of the slope, rather than simply assuming a uniform elliptical sliding surface. Therefore, there is still a pressing need to explore effective and accurate methods for evaluating the stability of regional 3D slopes under rainfall conditions.

As the complexity of slope stability increases with rainfall infiltration, traditional methods, including 2D analysis or simplified models, may fail to account for the full extent of the problem, leading to inaccurate results. This issue is particularly pronounced in regions with complex topography, heterogeneous material properties, and varying groundwater conditions, all of which can significantly impact the behavior of slopes under changing rainfall patterns. One promising tool for addressing these challenges is $FLAC^{3D}$, a powerful numerical modeling software that is widely used for simulating geotechnical systems. $FLAC^{3D}$ allows for the simulation of three-dimensional slope behavior, providing a more realistic representation of slope geometry and material heterogeneity than traditional 2D models. $FLAC^{3D}$ uses a finite difference method (FDM) to solve the governing equations of the system, allowing for the simulation of complex physical processes like soil-water interaction, seepage, and pore pressure distribution. This ability to model realistic groundwater flow and unsaturated soil behavior is crucial when simulating rainfall infiltration, as the redistribution of water within the slope can lead to significant changes in slope stability. In addition to its capability to model the physical processes of rainfall infiltration, $FLAC^{3D}$ also allows for the incorporation of a wide range of boundary conditions, such as surface loads, slope angles, and water table variations. These boundary conditions are critical for accurately capturing the real-world factors that influence slope stability, particularly in regions with variable rainfall patterns. In conclusion, $FLAC^{3D}$ offers a highly accurate and effective method for evaluating the stability of regional 3D slopes.

References

Althaus E, Friz-Töpfer A, Lempp C, Natau O (1994) Effects of water on strength and failure mode of coarse-grained granites at 300°C. Rock Mech Rock Eng 27:1–21. https://doi.org/10.1007/BF01025953

Amarasinghe MP, Kulathilaka SAS, Robert DJ et al (2024) Risk assessment and management of rainfall-induced landslides in tropical regions: a review. Nat Hazards 120:2179–2231. https://doi.org/10.1007/s11069-023-06277-3

Azizian A, Popescu R (2006) Three-dimensional seismic analysis of submarine slopes. Soil Dyn Earthq Eng 26:870–887. https://doi.org/10.1016/j.soildyn.2005.10.008

Brantut N, Heap MJ, Meredith PG, Baud P (2013) Time-dependent cracking and brittle creep in crustal rocks: a review. J Struct Geol 52:17–43. https://doi.org/10.1016/j.jsg.2013.03.007

Cai F, Ugai K (2004) Numerical analysis of rainfall effects on slope stability. Int J Geomech 4:69–78. https://doi.org/10.1061/(ASCE)1532-3641(2004)4:2(69)

Cascini L, Cuomo S, Pastor M, Sorbino G (2010) Modeling of rainfall-induced shallow landslides of the flow-type. J Geotech Geoenviron Eng 136:85–98. https://doi.org/10.1061/(ASCE)GT.1943-5606.0000182

Castagna A, Ougier-Simonin A, Benson PM et al (2018) Thermal damage and pore pressure effects of the brittle-ductile transition in Comiso limestone. J Geophys Res Solid Earth 123:7644–7660. https://doi.org/10.1029/2017JB015105

Chen H, Qin S, Xue L, Xu C (2021) Why the Xintan landslide was not triggered by the heaviest historical rainfall: mechanism and review. Eng Geol 294:106379. https://doi.org/10.1016/j.enggeo.2021.106379

Ching J, Wang J-S (2016) Application of the transitional Markov chain Monte Carlo algorithm to probabilistic site characterization. Eng Geol 203:151–167. https://doi.org/10.1016/j.enggeo.2015.10.015

Cho SE (2010) Probabilistic assessment of slope stability that considers the spatial variability of soil properties. J Geotech Geoenviron Eng 136:975–984. https://doi.org/10.1061/(ASCE)GT.1943-5606.0000309

Cho SE (2014) Probabilistic stability analysis of rainfall-induced landslides considering spatial variability of permeability. Engineering Geology 171:11–20. https://doi.org/10.1016/j.enggeo.2013.12.015

Cho SE, Lee SR (2001) Instability of unsaturated soil slopes due to infiltration. Comput Geotech 28:185–208. https://doi.org/10.1016/S0266-352X(00)00027-6

Cho MTT, Chueasamat A, Hori T et al (2021) Effectiveness of filter gabions against slope failure due to heavy rainfall. Soils Found 61:480–495. https://doi.org/10.1016/j.sandf.2021.01.010

Chueasamat A, Hori T, Saito H et al (2018) Experimental tests of slope failure due to rainfalls using 1g physical slope models. Soils Found 58:290–305. https://doi.org/10.1016/j.sandf.2018.02.003

Ciantia MO, Castellanza R, Crosta GB, Hueckel T (2015) Effects of mineral suspension and dissolution on strength and compressibility of soft carbonate rocks. Eng Geol 184:1–18. https://doi.org/10.1016/j.enggeo.2014.10.024

Danzer R, Supancic P, Pascual J, Lube T (2007) Fracture statistics of ceramics—Weibull statistics and deviations from Weibull statistics. Eng Fract Mech 74:2919–2932. https://doi.org/10.1016/j.engfracmech.2006.05.028

Dou H, Han T, Gong X, Zhang J (2014) Probabilistic slope stability analysis considering the variability of hydraulic conductivity under rainfall infiltration–redistribution conditions. Eng Geol 183:1–13. https://doi.org/10.1016/j.enggeo.2014.09.005

Duperret A, Taibi S, Mortimore RN, Daigneault M (2005) Effect of groundwater and sea weathering cycles on the strength of chalk rock from unstable coastal cliffs of NW France. Eng Geol 78:321–343. https://doi.org/10.1016/j.enggeo.2005.01.004

Erguler ZA, Ulusay R (2009) Water-induced variations in mechanical properties of clay-bearing rocks. Int J Rock Mech Min Sci 46:355–370. https://doi.org/10.1016/j.ijrmms.2008.07.002

Ersöz T, Topal T (2018) Assessment of rock slope stability with the effects of weathering and excavation by comparing deterministic methods and slope stability probability classification (SSPC). Environ Earth Sci 77:547. https://doi.org/10.1007/s12665-018-7728-4

Fang Z, Harrison JP (2002) Application of a local degradation model to the analysis of brittle fracture of laboratory scale rock specimens under triaxial conditions. Int J Rock Mech Min Sci 39:459–476. https://doi.org/10.1016/S1365-1609(02)00036-9

Froude MJ, and Petley DN (2018) Global fatal landslide occurrence from 2004 to 2016. Nat Hazards Earth Syst Sci 18: 2161–2181. https://doi.org/10.5194/nhess-18-2161-2018

Griffiths DV, Lane PA (1999) Slope stability analysis by finite elements. Géotechnique 49:387–403. https://doi.org/10.1680/geot.1999.49.3.387

Griffiths DV, Marquez RM (2007) Three-dimensional slope stability analysis by elasto-plastic finite elements. Géotechnique 57:537–546. https://doi.org/10.1680/geot.2007.57.6.537

Gu T, Wang J, Fu X, Liu Y (2015) GIS and limit equilibrium in the assessment of regional slope stability and mapping of landslide susceptibility. Bull Eng Geol Environ 74:1105–1115. https://doi.org/10.1007/s10064-014-0689-2

Gulino R, Phoenix SL (1991) Weibull strength statistics for graphite fibres measured from the break progression in a model graphite/glass/epoxy microcomposite. J Mater Sci 26:3107–3118. https://doi.org/10.1007/BF01124848

Gupta V, Bergström JS (1998) Compressive failure of rocks by shear faulting. J Geophys Res Solid Earth 103:23875–23895. https://doi.org/10.1029/97JB03107

Gutiérrez-Martín A (2020) A GIS-physically-based emergency methodology for predicting rainfall-induced shallow landslide zonation. Geomorphology 359:107121. https://doi.org/10.1016/j.geomorph.2020.107121

Hicks MA, Nuttall JD, Chen J (2014) Influence of heterogeneity on 3D slope reliability and failure consequence. Comput Geotech 61:198–208. https://doi.org/10.1016/j.compgeo.2014.05.004

Ho J-Y, Lee KT, Chang T-C et al (2012) Influences of spatial distribution of soil thickness on shallow landslide prediction. Eng Geol 124:38–46. https://doi.org/10.1016/j.enggeo.2011.09.013

Hoyos LR, Suescún-Florez EA, Puppala AJ (2015) Stiffness of intermediate unsaturated soil from simultaneous suction-controlled resonant column and bender element testing. Engineering Geology 188:10–28. https://doi.org/10.1016/j.enggeo.2015.01.014

Ietto F, Perri F, Cella F (2018) Weathering characterization for landslides modeling in granitoid rock masses of the capo Vaticano promontory (Calabria, Italy). Landslides 15:43–62. https://doi.org/10.1007/s10346-017-0860-5

Iverson RM (2000) Landslide triggering by rain infiltration. Water Resour Res 36:1897–1910. https://doi.org/10.1029/2000WR900090

Iyare UC, Blake OO, Ramsook R (2021) Estimating the uniaxial compressive strength of argillites using Brazilian tensile strength, ultrasonic wave velocities, and elastic properties. Rock Mech Rock Eng 54:2067–2078. https://doi.org/10.1007/s00603-020-02358-y

Iyare UC, Blake OO, Frash LP et al (2023) Water-weakening effects on the failure behavior of mudstones. Rock Mech Rock Eng 56:9171–9185. https://doi.org/10.1007/s00603-023-03544-4

Jaboyedoff M, Carrea D, Derron M-H et al (2020) A review of methods used to estimate initial landslide failure surface depths and volumes. Eng Geol 267:105478. https://doi.org/10.1016/j.enggeo.2020.105478

Jia N, Mitani Y, Xie M, Djamaluddin I (2012) Shallow landslide hazard assessment using a three-dimensional deterministic model in a mountainous area. Comput Geotech 45:1–10. https://doi.org/10.1016/j.compgeo.2012.04.007

Jia N, Mitani Y, Xie M et al (2015) GIS deterministic model-based 3D large-scale artificial slope stability analysis along a highway using a new slope unit division method. Nat Hazards 76:873–890. https://doi.org/10.1007/s11069-014-1524-6

Jiang S-H, Huang J, Griffiths DV, Deng Z-P (2022) Advances in reliability and risk analyses of slopes in spatially variable soils: a state-of-the-art review. Comput Geotech 141:104498. https://doi.org/10.1016/j.compgeo.2021.104498

Jiang Y, Chen L, Wang D et al (2024) Mechanical properties and acoustic emission characteristics of soft rock with different water contents under dynamic disturbance. Int J Coal Sci Technol 11:36. https://doi.org/10.1007/s40789-024-00682-0

Kasama K, Whittle AJ (2011) Bearing capacity of spatially random cohesive soil using numerical limit analyses. J Geotech Geoenviron Eng 137:989–996. https://doi.org/10.1061/(ASCE)GT.1943-5606.0000531

Kim DH, Gratchev I, Balasubramaniam A (2015) A photogrammetric approach for stability analysis of weathered rock slopes. Geotech Geol Eng 33:443–454. https://doi.org/10.1007/s10706-014-9830-z

Kim E, Stine MA, de Oliveira DBM, Changani H (2017) Correlations between the physical and mechanical properties of sandstones with changes of water content and loading rates. Int J Rock Mech Min Sci 100:255–262. https://doi.org/10.1016/j.ijrmms.2017.11.005

Kitamura R, Sako K (2010) Contribution of "soils and foundations" to studies on rainfall-induced slope failure. Soils Found 50:955–964. https://doi.org/10.3208/sandf.50.955

Lee Y-F, Chi Y-Y (2011) Rainfall-induced landslide risk at Lushan, Taiwan. Eng Geol 123: 113–121. https://doi.org/10.1016/j.enggeo.2011.03.006

Leong EC, Rahardjo H (2012) Two and three-dimensional slope stability reanalyses of Bukit Batok slope. Comput Geotech 42:81–88. https://doi.org/10.1016/j.compgeo.2012.01.001

Li D, Wong LNY, Liu G, Zhang X (2012) Influence of water content and anisotropy on the strength and deformability of low porosity meta-sedimentary rocks under triaxial compression. Eng Geol 126:46–66. https://doi.org/10.1016/j.enggeo.2011.12.009

Li L, Guan J, Liu Z (2021) A random discrete element method for modeling rock heterogeneity. Geomech Geophys Geoenerg Georesour 8:12. https://doi.org/10.1007/s40948-021-00320-y

Li X, Jiang Y, Sugimoto S (2024) Research on the influence of spatial dimensions on the stability of large-scale slopes under heavy rainfall. GeoHazards 5:732–754. https://doi.org/10.3390/geohazards5030037

Li X, Jiang Y, Sugimoto S (2025) Slope stability analysis under heavy rainfall considering the heterogeneity of weathered layers. Landslides 22:1257–1273. https://doi.org/10.1007/s10346-024-02404-8

Liakas S, O'Sullivan C, Saroglou C (2017) Influence of heterogeneity on rock strength and stiffness using discrete element method and parallel bond model. J Rock Mech Geotech Eng 9:575–584. https://doi.org/10.1016/j.jrmge.2017.02.003

Liu S, Huang X, Zhou A et al (2018) Soil-rock slope stability analysis by considering the nonuniformity of rocks. Math Probl Eng 2018:1–15. https://doi.org/10.1155/2018/3121604

Meng QX, Wang HL, Xu WY et al (2019) Multiscale strength reduction method for heterogeneous slope using hierarchical FEM/DEM modeling. Comput Geotech 115:103164. https://doi.org/10.1016/j.compgeo.2019.103164

Miščević P, Vlastelica G (2014) Impact of weathering on slope stability in soft rock mass. J Rock Mech Geotech Eng 6:240–250. https://doi.org/10.1016/j.jrmge.2014.03.006

Ng CWW, Shi Q (1998) A numerical investigation of the stability of unsaturated soil slopes subjected to transient seepage. Comput Geotech 22:1–28. https://doi.org/10.1016/S0266-352X(97)00036-0

Noël C, Baud P, Violay M (2021) Effect of water on sandstone's fracture toughness and frictional parameters: brittle strength constraints. Int J Rock Mech Min Sci 147:104916. https://doi.org/10.1016/j.ijrmms.2021.104916

Oh S, Lu N (2015) Slope stability analysis under unsaturated conditions: case studies of rainfall-induced failure of cut slopes. Eng Geol 184:96–103. https://doi.org/10.1016/j.enggeo.2014.11.007

Olona J, Pulgar JA, Fernández-Viejo G et al (2010) Weathering variations in a granitic massif and related geotechnical properties through seismic and electrical resistivity methods. Near Surf Geophys 8:585–599. https://doi.org/10.3997/1873-0604.2010043

Pan Y, Wu G, Zhao Z, He L (2020) Analysis of rock slope stability under rainfall conditions considering the water-induced weakening of rock. Comput Geotech 128:103806. https://doi.org/10.1016/j.compgeo.2020.103806

Park HJ, Lee JH, Woo I (2013) Assessment of rainfall-induced shallow landslide susceptibility using a GIS-based probabilistic approach. Eng Geol 161:1–15. https://doi.org/10.1016/j.enggeo.2013.04.011

Perri F, Ietto F, Le Pera E, Apollaro C (2016) Weathering processes affecting granitoid profiles of capo Vaticano (Calabria, southern Italy) based on petrographic, mineralogic and reaction path modelling approaches. Geol J 51:368–386. https://doi.org/10.1002/gj.2635

Rafiei Renani H, Martin CD (2020) Factor of safety of strain-softening slopes. J Rock Mech Geotech Eng 12:473–483. https://doi.org/10.1016/j.jrmge.2019.11.004

Sabzevari T, Noroozpour S (2014) Effects of hillslope geometry on surface and subsurface flows. Hydrgeol J 22:1593–1604. https://doi.org/10.1007/s10040-014-1149-6

Sheikh MR, Nakata Y, Shitano M, Kaneko M (2021) Rainfall-induced unstable slope monitoring and early warning through tilt sensors. Soils Found 61:1033–1053. https://doi.org/10.1016/j.sandf.2021.05.010

Singh MR, Bhargava SC (1994) Magnetic behavior of Ga0.3Fe0.7CuCrO4. Solid State Commun 90:183–187. https://doi.org/10.1016/0038-1098(94)90338-7

Srivastava A, Babu GLS, Haldar S (2010) Influence of spatial variability of permeability property on steady state seepage flow and slope stability analysis. Eng Geol 110:93–101. https://doi.org/10.1016/j.enggeo.2009.11.006

Suchomel R, Mašı'n D (2010) Comparison of different probabilistic methods for predicting stability of a slope in spatially variable c–φ soil. Comput Geotech 37:132–140. https://doi.org/10.1016/j.compgeo.2009.08.005

Sun C, Chai J, Xu Z, Qin Y (2017) 3D stability charts for convex and concave slopes in plan view with homogeneous soil based on the strength-reduction method. Int J Geomech 17:06016034. https://doi.org/10.1061/(ASCE)GM.1943-5622.0000809

Talebi A, Troch PA, Uijlenhoet R (2008) A steady-state analytical slope stability model for complex hillslopes. Hydrol Process 22:546–553. https://doi.org/10.1002/hyp.6881

Tang SB, Tang CA (2012) Numerical studies on tunnel floor heave in swelling ground under humid conditions. Int J Rock Mech Min Sci 55:139–150. https://doi.org/10.1016/j.ijrmms.2012.07.007

Tang CA, Liu H, Lee PKK et al (2000) Numerical studies of the influence of microstructure on rock failure in uniaxial compression—part I: effect of heterogeneity. Int J Rock Mech Min Sci 37:555–569. https://doi.org/10.1016/S1365-1609(99)00121-5

Tang S, Tang C, Liang Z et al (2012) Numerical study of the influence of material structure on effective thermal conductivity of concrete. Heat Transfer Eng 33:732–747. https://doi.org/10.1080/01457632.2011.635988

Tang SB, Zhang H, Tang CA, Liu HY (2016) Numerical model for the cracking behavior of heterogeneous brittle solids subjected to thermal shock. Int J Solids Struct 80:520–531. https://doi.org/10.1016/j.ijsolstr.2015.10.012

Tang S, Yu C, Tang C (2018) Numerical modeling of the time-dependent development of the damage zone around a tunnel under high humidity conditions. Tunn Undergr Space Technol 76:48–63. https://doi.org/10.1016/j.tust.2018.03.012

Tran TV, Alvioli M, Lee G, An HU (2017) Three-dimensional, time-dependent modeling of rainfall-induced landslides over a digital landscape: a case study. Landslides 15:1071–1084. https://doi.org/10.1007/s10346-017-0931-7

Tran TV, Alvioli M, Hoang VH (2022) Description of a complex, rainfall-induced landslide within a multi-stage three-dimensional model. Nat Hazards 110:1953–1968. https://doi.org/10.1007/s11069-021-05020-0

Tschuchnigg F, Schweiger HF, Sloan SW (2015) Slope stability analysis by means of finite element limit analysis and finite element strength reduction techniques. Part I: numerical studies considering non-associated plasticity. Comput Geotech 70:169–177. https://doi.org/10.1016/j.compgeo.2015.06.018

Van Den Bout B, Lombardo L, Chiyang M et al (2021) Physically-based catchment-scale prediction of slope failure volume and geometry. Eng Geol 284:105942. https://doi.org/10.1016/j.enggeo.2020.105942

van Mier JGM, van Vliet MRA, Wang TK (2002) Fracture mechanisms in particle composites: statistical aspects in lattice type analysis. Mech Mater 34:705–724. https://doi.org/10.1016/S0167-6636(02)00170-9

Van Westen CJ, Van Asch TWJ, Soeters R (2006) Landslide hazard and risk zonation—why is it still so difficult? Bull Eng Geol Environ 65:167–184. https://doi.org/10.1007/s10064-005-0023-0

Vanmarcke E, Grigoriu M (1983) Stochastic finite element analysis of simple beams. J Eng Mech 109:1203–1214. https://doi.org/10.1061/(ASCE)0733-9399(1983)109:5(1203)

Vásárhelyi B, Ván P (2006) Influence of water content on the strength of rock. Eng Geol 84:70–74. https://doi.org/10.1016/j.enggeo.2005.11.011

Wang J, Chen L (2021) The effect of hillslope geometry on Hortonian rainfall-infiltration-runoff processes. J Hydrol 594:125962. https://doi.org/10.1016/j.jhydrol.2021.125962

Wang D, Wang B, Yuan W, Liu L (2023) Investigation of rainfall intensity on the slope failure process using GPU-accelerated coupled MPM. Comput Geotech 163:105718. https://doi.org/10.1016/j.compgeo.2023.105718

Wasantha PLP, Ranjith PG (2014) Water-weakening behavior of Hawkesbury sandstone in brittle regime. Eng Geol 178:91–101. https://doi.org/10.1016/j.enggeo.2014.05.015

Wei WB, Cheng YM, Li L (2009) Three-dimensional slope failure analysis by the strength reduction and limit equilibrium methods. Comput Geotech 36:70–80. https://doi.org/10.1016/j.compgeo.2008.03.003

Weibull W (1939) A statistical theory of the strength of materials. Proceedings of the American Mathematical Society 151(5):1034.

Wines D (2016) A comparison of slope stability analyses in two and three dimensions. J South Afr Inst Min Metall 116:399–406. https://doi.org/10.17159/2411-9717/2016/v116n5a5

Xie M, Esaki T, Cai M (2004) A GIS-based method for locating the critical 3D slip surface in a slope. Comput Geotech 31:267–277. https://doi.org/10.1016/j.compgeo.2004.03.003

Xie M, Tetsuro E, Qiu C, Jia L (2007) Spatial three-dimensional landslide susceptibility mapping tool and its applications. Earth Sci Front 14:73–84. https://doi.org/10.1016/S1872-5791(08)60004-4

Xiong X, Shi Z, Xiong Y et al (2019) Unsaturated slope stability around the three gorges reservoir under various combinations of rainfall and water level fluctuation. Eng Geol 261:105231. https://doi.org/10.1016/j.enggeo.2019.105231

Xu W, Yue Z, Hu R (2008) Study on the mesostructure and mesomechanical characteristics of the soil–rock mixture using digital image processing based finite element method. Int J Rock Mech Min Sci 45:749–762. https://doi.org/10.1016/j.ijrmms.2007.09.003

Xu Y, Li J, Fan H et al (2017) Stability analysis of clastic rock slope with mudstone interlayer under rainfall infiltration. Geotech Geol Eng 35:1871–1883. https://doi.org/10.1007/s10706-017-0215-y

Xu J, Zhao X, Li P, Zhang M (2021) Stability of a 3D unsaturated vertical cut slope subjected to variable rainfall infiltration. Comput Geotech 134:104110. https://doi.org/10.1016/j.compgeo.2021.104110

Yilmaz I (2010) Influence of water content on the strength and deformability of gypsum. Int J Rock Mech Min Sci 47:342–347. https://doi.org/10.1016/j.ijrmms.2009.09.002

Yu C, Tang S, Tang C et al (2019) The effect of water on the creep behavior of red sandstone. Eng Geol 253:64–74. https://doi.org/10.1016/j.enggeo.2019.03.016

Zhang Y, Chen G, Zheng L et al (2013) Effects of geometries on three-dimensional slope stability. Can Geotech J 50:233–249. https://doi.org/10.1139/cgj-2012-0279

Zhang LL, Fredlund MD, Fredlund DG et al (2015) The influence of the unsaturated soil zone on 2-D and 3-D slope stability analyses. Eng Geol 193:374–383. https://doi.org/10.1016/j.enggeo.2015.05.011

Zhang Q, Wang L, Zhang H (2022) Rainfall infiltration process of a rock slope with considering the heterogeneity of saturated hydraulic conductivity. Front Earth Sci 9:804005. https://doi.org/10.3389/feart.2021.804005

Zhang C, Bai Q, Han P et al (2023) Strength weakening and its micromechanism in water–rock interaction, a short review in laboratory tests. Int J Coal Sci Technol 10:10. https://doi.org/10.1007/s40789-023-00569-6

Zhao L, Yang F, Zhang Y et al (2015) Effects of shear strength reduction strategies on safety factor of homogeneous slope based on a general nonlinear failure criterion. Comput Geotech 63:215–228. https://doi.org/10.1016/j.compgeo.2014.08.015

Zhao Z, Guo T, Ning Z et al (2018) Numerical modeling of stability of fractured reservoir bank slopes subjected to water–rock interactions. Rock Mech Rock Eng 51:2517–2531. https://doi.org/10.1007/s00603-017-1360-6

Zhou Y, Qi S-C, Fan G et al (2020) Topographic effects on three-dimensional slope stability for fluctuating water conditions using numerical analysis. Water 12:615. https://doi.org/10.3390/w12020615

Zienkiewicz OC, Humpheson C, Lewis RW (1975) Associated and non-associated visco-plasticity and plasticity in soil mechanics. Géotechnique 25:671–689. https://doi.org/10.1680/geot.1975.25.4.671

Chapter 2
Laboratory Tests on Rainfall Infiltration in Slopes

Abstract This chapter focuses on laboratory experiments for measuring the hydraulic parameters of unsaturated soils. These parameters include the soil–water characteristic curve (SWCC) and the permeability coefficient for unsaturated soils, both of which play a critical role in evaluating slope stability during rainfall events. Their importance lies in providing the fundamental input data for seepage and stability analyses. The chapter then presents a one-dimensional seepage test, accompanied by a numerical seepage analysis conducted under identical boundary conditions to validate and interpret the experimental results. It also describes a rainfall-induced slope model test, designed to simulate the coupled hydrological and mechanical responses of slopes under rainfall infiltration. Corresponding numerical simulations are performed to enable a comparative analysis, thereby enhancing the understanding of slope deformation and failure mechanisms. These experimental and numerical investigations offer valuable insights into the hydrological and geotechnical processes that govern slope stability under rainfall conditions.

Keywords Laboratory experiments · Permeability coefficient · SWCC · Seepage analysis · Fluid–solid coupling

2.1 Hydraulic Soil Parameters

According to the Japan Society of Civil Engineers (JSCE), rainfall in slopes can be categorized into surface runoff and infiltration water, as illustrated in Fig. 2.1. The unsaturated zone lies between the ground surface and the groundwater table, which is characterized by negative pore pressure. Furthermore, factors such as variations in soil types within the slope, groundwater supply from surrounding boundaries, vegetation cover, and evaporation significantly influence the infiltration dynamics.

The movement of water within unsaturated soil is governed by the chemical potential of water ψ_T. It comprises several components including gravitational potential ψ_g, determined by elevation, osmotic potential ψ_{sol}, driven by osmotic

Y. Jiang, X. Li, *Stability Evaluation of Slope Under Heavy Rainfall Based on Geometric Features*, Springer Natural Hazards,
https://doi.org/10.1007/978-981-95-8656-1_2

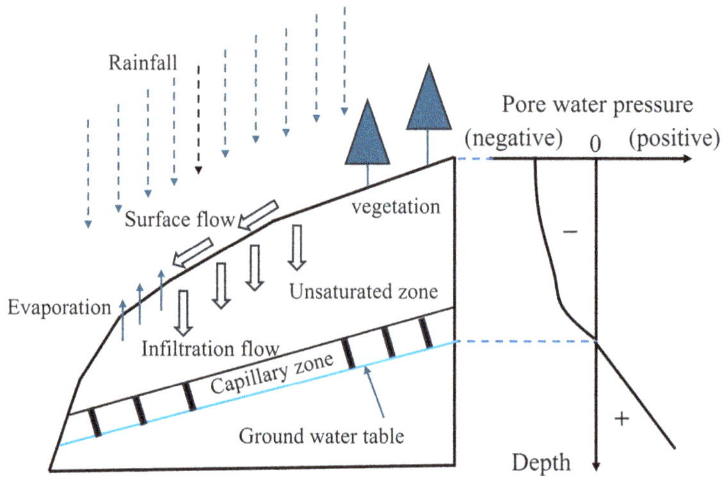

Fig. 2.1 Rainwater infiltration into a slope

pressure differences due to ion concentration, matric potential ψ_m, which is influenced by soil adsorption forces and surface tension, and air pressure potential ψ_a. Under atmospheric pressure, the air pressure potential is typically zero and is often included in the matric potential during experimental conditions. Both water and air coexist in the voids of unsaturated soils. Due to surface tension, a pressure difference arises between pore air pressure and pore water pressure, referred to as suction. Suction increases the interparticle adhesion forces in unsaturated soils, enhancing the resistance to particle slippage at contact points. As a result, the strength of unsaturated soil is greater than that of saturated soil. Figure 2.2 shows the presence of a meniscus at the contact point in the two-sphere model, which assumes the soil particles to be of equal size. In this case, the magnitude of the particle-to-particle adhesion force N_c caused by the meniscus is expressed by the following equation:

$$N_c = 2\pi \cdot r_2 \cdot T_s + \pi \cdot r_2^2 \cdot s \tag{2.1}$$

where N_c represents the interparticle adhesion force, T_s represents the surface tension of water, and r_2 represents the meniscus radius. The first term on the right-hand side of Eq. (2.1) represents the effect of surface tension, while the second term represents the effect of suction on the interparticle adhesion force.

This explanation refers to a two-sphere model used in soil mechanics or granular materials, where the adhesion force between two particles is influenced by the presence of a meniscus (a thin film of liquid) at the contact point. The equation for the adhesion force would typically consider factors such as the radius of the spheres, surface tension, and the characteristics of the liquid forming the meniscus.

The conceptual diagram of the soil-water characteristic curve (SWCC), which represents the relationship between suction and water content, is illustrated in Fig. 2.3. Water content can be expressed using parameters such as the degree of

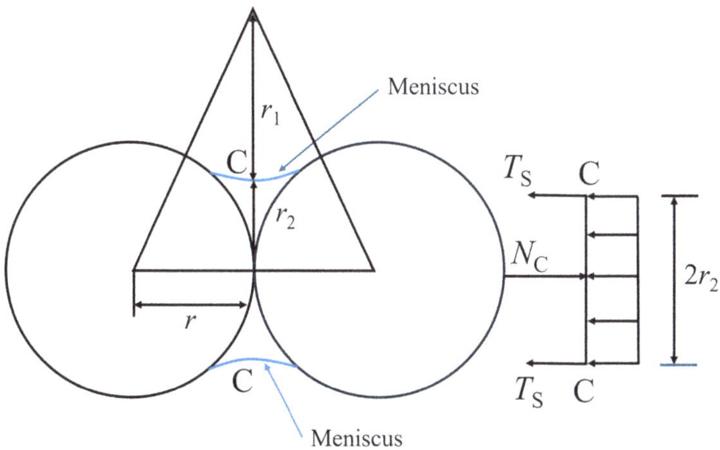

Fig. 2.2 Particle-to-particle adhesion force using the two-sphere model

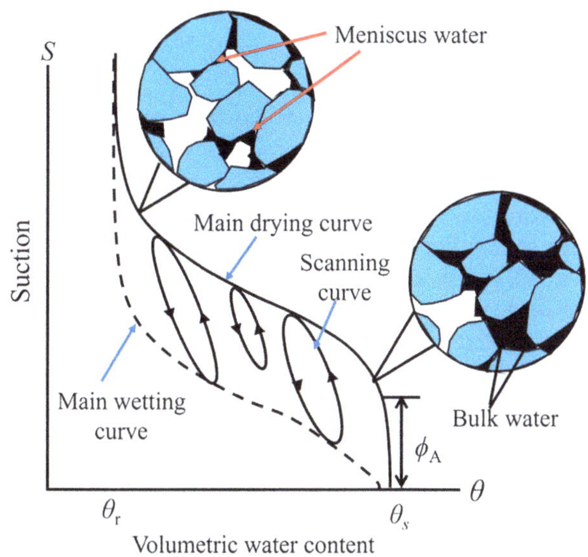

Fig. 2.3 Soil water characteristic curve and moisture condition

saturation S_r, the water content ratio w, and the volumetric water content θ. The solid line represents the main drying curve, while the dashed line represents the main wetting curve. θ_r is the residual water content (or residual moisture content). This figure illustrates the relationship between soil suction (or pressure head) and moisture content, showing how moisture changes in response to varying levels of suction, typically in the context of unsaturated soil behavior. The curve is essential for understanding the behavior of water in soils, influencing processes such as infiltration, retention, and drainage.

As shown in Fig. 2.3, the relationship between suction and water content is not univocal, it exhibits hysteresis during the drying and wetting processes. This hysteresis arises from several factors, including the non-uniform geometric configuration of pores, differences in the contact angle of pore water with soil particles during the advancement or retreat within the pores, the presence of entrapped air within the pores (commonly referred to as "entrapped air"), and structural changes in the soil skeleton. These factors collectively contribute to the distinct paths observed during the drying and wetting processes in the SWCC. The unsaturated hydraulic conductivity is defined by extending Darcy's law to unsaturated flow fields, using the total head difference, which comprises the pressure head (the difference between pore air pressure and pore water pressure) and the elevation head. Since unsaturated hydraulic conductivity is highly dependent on the water retention properties of the soil, its hysteresis with respect to water content is negligible during the drying and wetting processes. However, hysteresis cannot be ignored when considering suction. Consequently, the unsaturated hydraulic conductivity K_r is often expressed as a ratio of the saturated hydraulic conductivity k_s, as shown in Eq. (2.2).

$$K_r = \frac{k(\theta)}{k_s} \tag{2.2}$$

Research using various analytical methods has been conducted to understand the slope failure mechanisms induced by rainfall infiltration. For example, Wei et al. (2009) have investigated methods for analyzing slope stability using limit equilibrium analysis. During heavy rainfall events, slope failure is believed to occur due to an increase in the sliding force caused by the rise in the saturation of soils, as well as an increase in pore pressure and a reduction in suction. This process weakens the shear resistance of the soil and results in slope instability. Therefore, accurately understanding the infiltration behavior of rainwater and the changes in pore water pressure is essential for evaluating slope stability during heavy rainfall events. Rainfall infiltrates through the unsaturated zone of the soil slope, which necessitates considering the mechanics of unsaturated soil behavior in slope stability analysis. As the relationship between the unsaturated soil's permeability and the volumetric water content, and the soil water characteristic curve between volumetric water content and suction are nonlinear, it is challenging to account for these factors using simplified numerical methods. Therefore, Cai and Ugai (2004) conducted a saturated-unsaturated infiltration analysis using the finite element method. This approach calculates the amount of rainwater infiltrating within the slope and is a powerful tool for understanding the temporal changes in pore water pressure during heavy rainfall. In natural slopes, the saturation at the slope surface may increase during heavy rainfall. If the free movement of air is obstructed, air may become trapped within the pores, leading to an increase in air pressure. This elevated air pressure could contribute to slope failure. Therefore, improving the reproducibility of analyses that take both pore water pressure and pore air pressure into account will significantly aid in understanding the mechanisms behind slope failure.

Table 2.1 The mechanical properties of remolded soils

Parameters	Value
Moisture content (%)	10
Compaction energy (kJ/m^3)	336.9
Dry density (kg/m^3)	1500
Porosity	0.43
Permeability (m/s)	1.0×10^{-6}

The determination of hydraulic soil parameters used in numerical simulations is crucial for ensuring the accuracy of the calculation results. Therefore, using the Sasebo landslide as a case study, this study utilized residual soil samples collected from the Sasebo landslide to analyze the Soil-Water Characteristic Curve (SWCC) and permeability. The soil sample is prepared by compaction to achieve the specified wet density. The physical properties of the remolded soil are presented in Table 2.1. A sample with a volume of 1000 cm^3 was compacted in this manner and a permeability test (JGS0311) was conducted to determine the unsaturated and saturated permeability of the remold soil. The permeability coefficient k in the constant head permeability test is given as follows:

$$k = \frac{QL}{\Delta hAt} \tag{2.3}$$

where Q represents the fluid flow rate, k is the permeability coefficient, A is the flow area, L is the flow path length, Δh is the hydraulic head difference, and t is the time.

The van Genuchten (VG) model (1980) is widely used to fit the soil-water characteristic curve (SWCC). It can effectively capture the hydraulic behavior of unsaturated soils (Kafle et al. 2022). In the experiment (Fig. 2.4), two tensiometers and two soil moisture sensors were installed. Each tensiometer recorded two datasets, resulting in a total of four datasets (Li et al. 2025). The measurements were averaged to ensure reliability. Incremental water addition allowed a series of consecutive readings to be obtained (Fig. 2.5). The VG model can be expressed as follows:

$$S_e = \left[\frac{1}{1+(as)^n}\right]^{1-\frac{1}{n}} \tag{2.4}$$

where S_e is the effective degree of saturation and s is the matric suction. In this study, matric suction is expressed as negative pore pressure based on single-phase analysis. The parameters a and n are empirical constants that characterize the hydraulic behavior of unsaturated soils.

By plotting volumetric moisture content against matric suction, the relationship between soil water content and water pressure can be characterized using the van Genuchten (VG) functions. As shown in Fig. 2.6, the parameters were determined by fitting the soil moisture characteristic curve to the VG model ($a = 0.125$, $n = 1.49$). The corresponding relationship between degree of saturation and matric suction can

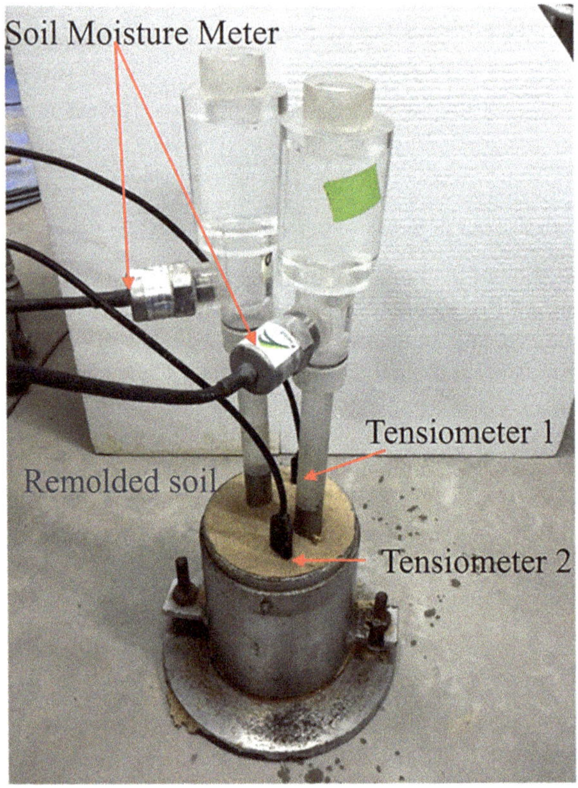

Fig. 2.4 Measurement of matric suction and soil moisture content (Li et al. 2025)

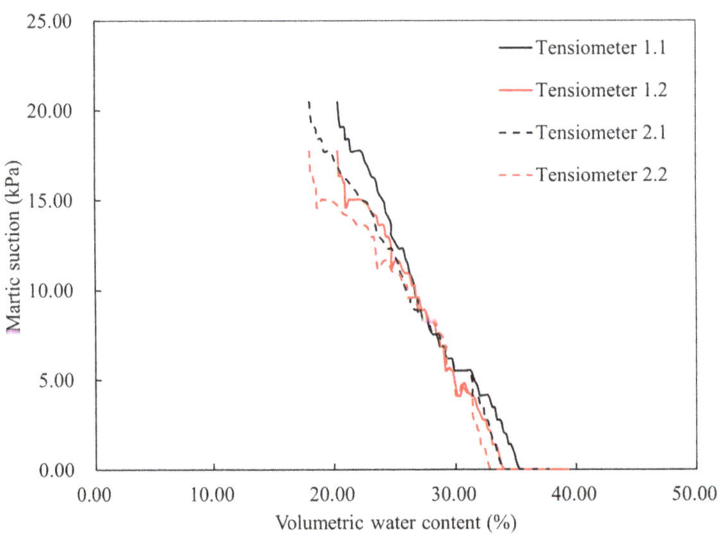

Fig. 2.5 The relationship between volumetric water content and matric suction (Li et al. 2025)

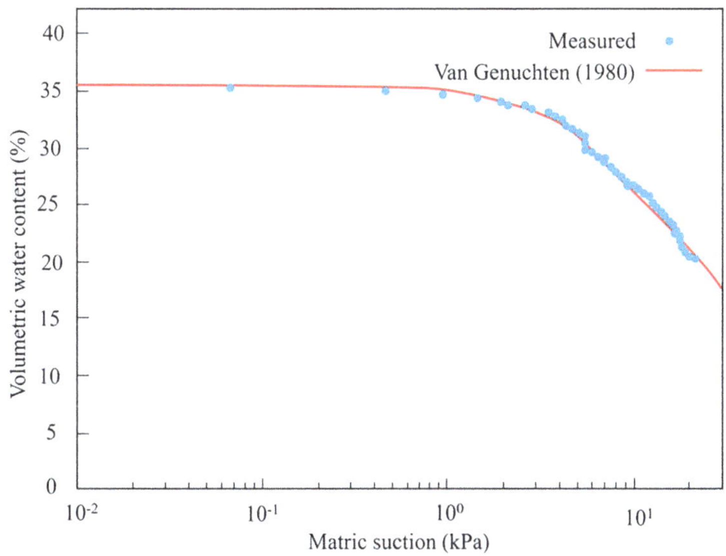

Fig. 2.6 The soil-water characteristic curve obtained from the nonlinear regression according to Eq. (2.4) (Li et al. 2025)

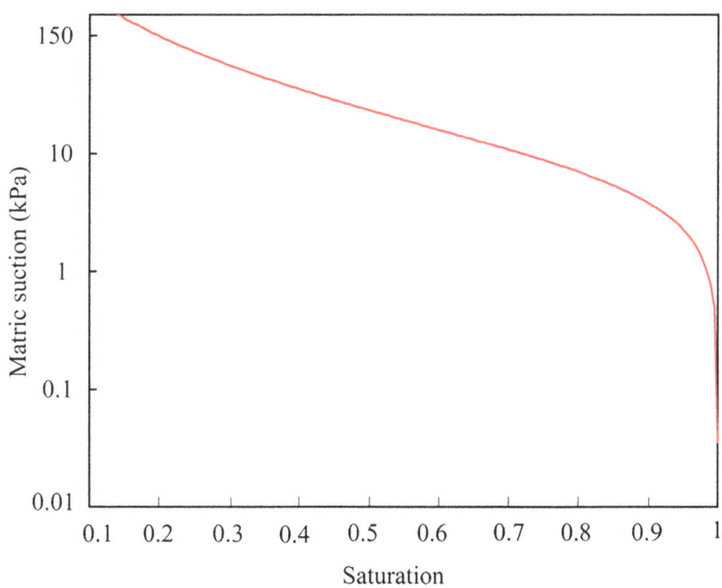

Fig. 2.7 The soil-water characteristic curve of the test sample ($a = 0.125$, $n = 1.49$) (Li et al. 2025)

be obtained through a straightforward conversion from volumetric moisture content (Fig. 2.7). Using the SWCC in the 0–20 kPa range, the maximum suction at the slope surface prior to rainfall was estimated to be approximately 100–150 kPa (Li et al. 2025).

2.2 One-Dimensional (1D) Seepage Analysis

A one-dimensional unsaturated seepage experiment was conducted using the same experimental apparatus for measuring permeability, as depicted in Fig. 2.8a. The specimen has a cylindrical shape with a diameter of 280 mm and a height of 500 mm. Positive pore water pressure and negative pore water pressure are measured at two locations using piezometers and tensiometers: the central part and the lower part of the specimen, respectively (Fig. 2.8a). When preparing samples for a one-dimensional permeation flow model experiment, the compaction effort applied to the samples is equivalent to the compaction energy used in actual engineering practices. Figure 2.8b shows the numerical model identical to the experimental model, with two pore pressure monitoring points set at the same locations. The results of the model experiment show that the pore pressure at measuring point No. 1, located in the central part of the model, begins to increase first, followed by an increase in the pore pressure at the lower measuring point No. 2 (Fig. 2.9). This confirms that the infiltration front caused by water infiltration from the upper part progressed downward. Approximately 40 min later, the pore pressure at measuring point 1 began to change from negative to positive, indicating that the wetting front had progressed from top to bottom and reached measuring point 1. About 70 min later, the pore pressure at measuring point 2 started to change from negative to positive, meaning that the wetting front had advanced to measuring point 2. Approximately 250 min after the start of water supply, the pore pressure at measuring points No.1 and No. 2 began to increase to positive values (Fig. 2.9). This is due to the saturation front rising from the lower part of the model after the leading infiltration front reaches the bottom. Subsequently, both measuring points converged to values equivalent to the hydrostatic pressure when the entire model became saturated. Based on the soil water retention curve obtained from the experiment, a

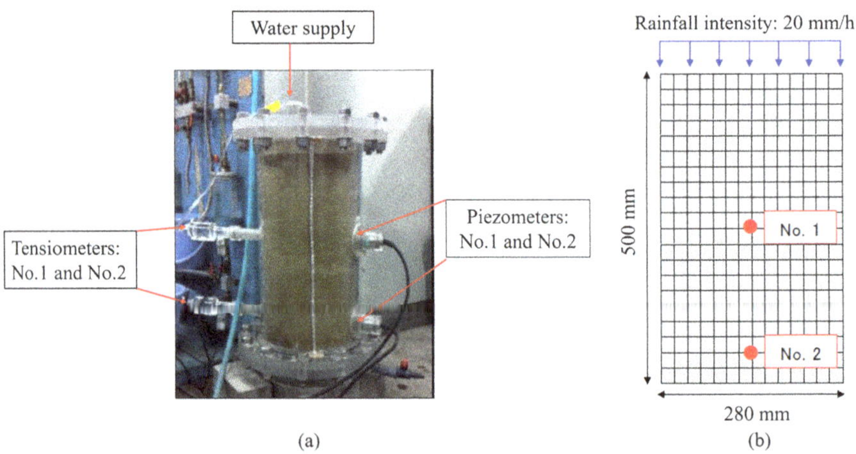

Fig. 2.8 (**a**) One-dimensional model test for permeation flow analysis and (**b**) One-dimensional numerical model for permeation flow analysis

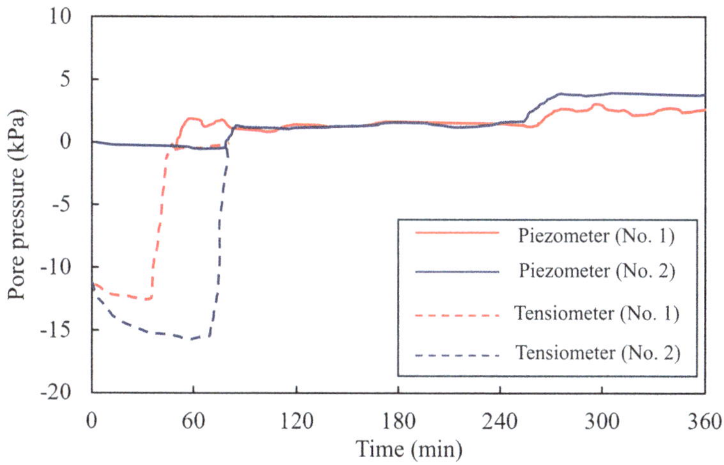

Fig. 2.9 Results of laboratory tests on pore pressure variation

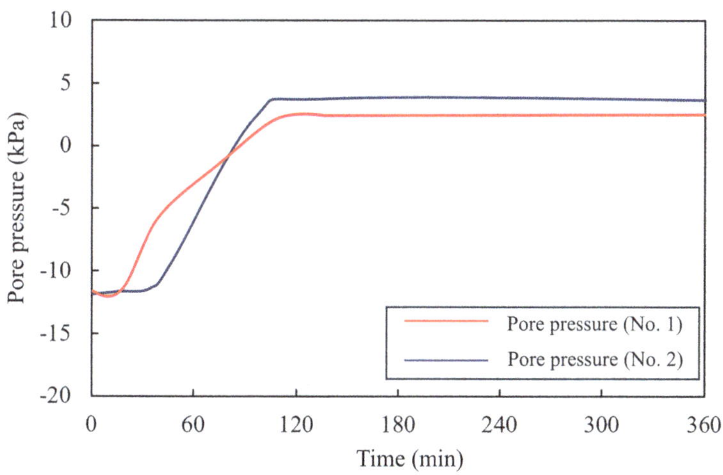

Fig. 2.10 Results of numerical simulations on pore pressure variation

one-dimensional seepage numerical simulation was conducted to verify the accuracy of the VG model fitting parameters. Measurement points No. 1 and No. 2 were placed at the same positions, as illustrated in Fig. 2.8b. The results of the numerical simulation are shown in Fig. 2.10, where it can be observed that the pore pressure variations at the measurement points from both the experiment and simulation are quite similar. It is important to note that the permeability of the soil in the simulation is assumed to be isotropic, while the soil permeability is anisotropic in the experiment. Therefore, the difference in pore pressure variations resulting from the disparity in permeability coefficients is reasonable. This also supports the accuracy of the empirical parameters a and n in the fitted soil-water characteristic curve.

In traditional seepage analysis, it is assumed that the region above the free surface is dry, and there is no seepage at the free surface. However, with the development and refinement of the saturation-unsaturation theory, the beneficial effects of negative pore water pressure in the unsaturated zone on slopes, dams, and other engineering structures have been fully recognized (Ng and Shi 1998; Pan et al. 2020). FLAC^{3D} is an internationally recognized three-dimensional finite difference software for geomechanical analysis. It offers a secondary development platform based on the FISH programming language and is particularly proficient in solving fluid-solid coupling problems. In FLAC^{3D}, however, it is not possible to directly solve for the free surface and overflow points of seepage, and the permeability coefficient can only be set as a fixed constant. To overcome the above challenges, this study uses FISH functions for secondary development of seepage module, applying a modified permeability coefficient method to solve for the free surface and overflow location of seepage. Based on this method, the characteristics of unsaturated soils are used to iteratively modify the permeability coefficient of the unsaturated zone, thereby calculating a more accurate saturated-unsaturated seepage field. According to the obtained soil-water characteristic curve, this section introduces the significant impact of real-time correction of the permeability coefficient of unsaturated soils on the accuracy of numerical simulation results during rainfall infiltration process. The governing equation for seepage in unsaturated soils, in the case of isotropic and homogeneous conditions, is expressed as follows:

$$\frac{\partial}{\partial x}\left(k_{\mathrm{w}}\frac{\partial h}{\partial x}\right)+\frac{\partial}{\partial y}\left(k_{\mathrm{w}}\frac{\partial h}{\partial y}\right)+\frac{\partial}{\partial z}\left(k_{\mathrm{w}}\frac{\partial h}{\partial z}\right)=n\frac{\partial S_{\mathrm{w}}}{\partial t}+S_{\mathrm{w}}S_{\mathrm{s}}\frac{\partial h}{\partial t} \tag{2.5}$$

where h represents the hydraulic head, the permeability coefficient k_{w} is a function of volumetric water content n_{w}, S_{w} represents the saturation, and S_{s} denotes the unit storage.

To solve the saturated-unsaturated seepage field, the most critical aspect is determining the interface between the saturated and unsaturated zones, known as the free surface. Eq. (2.5) is the general differential equation used for saturated-unsaturated seepage calculations. Neuman proposed a mathematical model that considers both saturated and unsaturated conditions for numerical calculations. In this equation, the hydraulic head h can be replaced by the pressure head p, and the volumetric water content n_{w} can replace the saturation S_{w}. By defining the specific water capacity C as:

$$C=\frac{\partial n_{\mathrm{w}}}{\partial p} \tag{2.6}$$

Equation (2.5) can be rewritten as follows:

$$\frac{\partial}{\partial x}\left(k_{\mathrm{w}}\frac{\partial p}{\partial x}\right)+\frac{\partial}{\partial y}\left(k_{\mathrm{w}}\frac{\partial p}{\partial y}\right)+\frac{\partial}{\partial z}\left(k_{\mathrm{w}}\frac{\partial p}{\partial z}\right)+\frac{\partial k_{\mathrm{w}}}{\partial z}=\left(C+\frac{n_{\mathrm{w}}}{n}S_{\mathrm{s}}\right)\frac{\partial p}{\partial t} \tag{2.7}$$

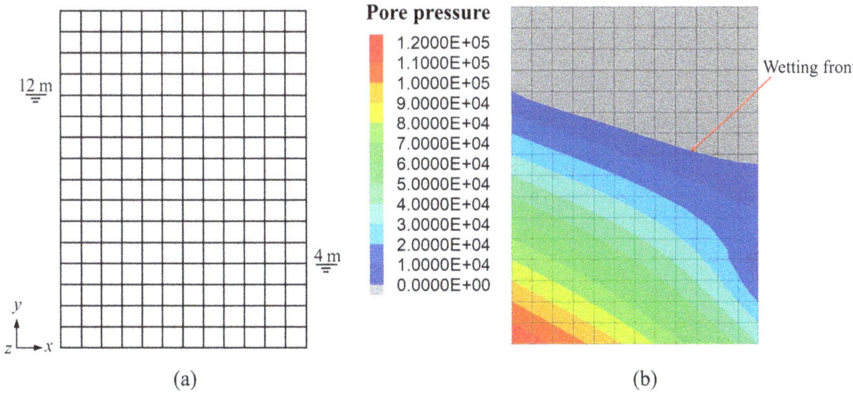

Fig. 2.11 (**a**) Calculation model and (**b**) Seepage field only defined upstream and downstream heads

In Eq. (2.7), the pressure head p is a function of the volumetric water content n_w. With known initial and boundary conditions, the above equation can be solved. The points where the computed $p = 0$ represent the free surface of seepage field. This study employs a model measuring 12 m in width, 16 m in height, and 1 m in thickness (Fig. 2.11a). The upstream hydraulic head is set to 12 m, and the downstream hydraulic head is set to 4 m. The saturated permeability coefficient is assumed to be 1×10^{-7} m/s. The model is divided into 192 grids ($12 \times 16 \times 1$) as shown in Fig. 2.11a. Boundary conditions for the hydraulic head are defined at the upstream and downstream ends, and the permeability coefficient for the entire region is assumed to be saturated for the calculations. The isoline where the pore water pressure equals zero is identified as the initial wetting front, as shown in Fig. 2.11b. The area above this line represents the unsaturated zone, while the area below it represents the saturated zone.

The regions where the pore water pressure is less than 0 are identified as the unsaturated zone using FISH functions. The permeability coefficient of the saturated zone remains unchanged, while the permeability coefficient of the unsaturated zone is reassigned as $k' = 1 \times 10^{-18}$. This ensures that the flow exchange between the saturated and unsaturated zones is minimal and can be neglected. During the iterative process, the permeability coefficients are repeatedly adjusted as described. If the number of steps in the next iteration is fewer than 10, the computation stops, as the difference in hydraulic head at each point between consecutive iterations becomes negligible. As shown in Fig. 2.12a, the isoline of zero pore water pressure (phreatic line) delineates the unsaturated zone above it. In this region, the permeability coefficient is so small that there is virtually no seepage. This iterative process produces a curved phreatic line. The curvature is caused by the abrupt change in permeability coefficients between the two zones, resulting in local refraction of water flow. However, based on the principle that the free surface within any homogeneous medium must be continuous, smooth, and free of back-turning, this phreatic line requires further correction.

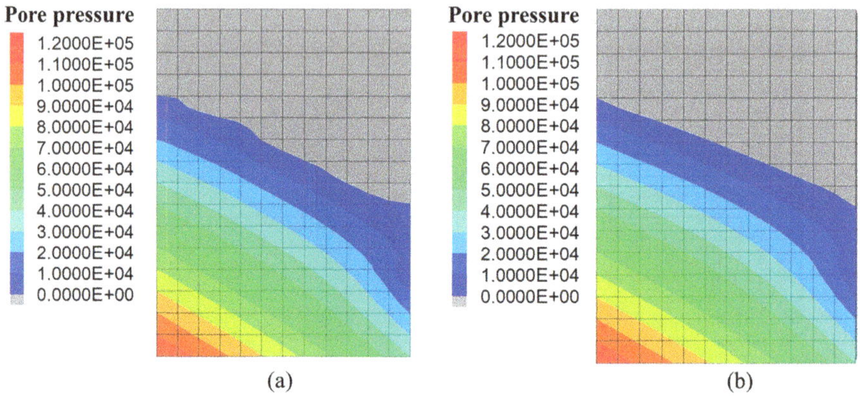

Fig. 2.12 (**a**) Seepage field after adjusting permeability, and (**b**) modified saturated-unsaturated seepage field (Pa)

Building upon the preceding calculations, the nodes on the downstream overflow boundary are recorded, and their pore water pressures are reassigned to 0. From the diagram, it can be roughly observed that the overflow boundary ranges from $z = 4$ m to $z = 7$ m. Next, using the relationship between permeability coefficient and matric suction, the permeability coefficient of the unsaturated zone is modified again, and iterative calculations are performed. During the iterative process, the steps of adjusting the permeability coefficient are repeated. If the number of steps in the next iteration is fewer than 10, the calculation stops. The final results demonstrate a very smooth phreatic line without any back-turning, as shown in Fig. 2.12b. This is because the transition of the permeability coefficient from the saturated zone to the unsaturated zone is a gradual process rather than a sudden jump. This method successfully achieves a smooth transition that aligns well with actual conditions. The method of modifying the permeability coefficient was implemented to identify the phreatic surface and overflow points. Additionally, the phreatic surface was refined based on the functional relationship between the permeability coefficient and pore water pressure, ensuring a smooth surface without reversals, and yielding a comprehensive saturated-unsaturated seepage field. This demonstrates the accuracy of FLAC^{3D} in unsaturated seepage analysis, highlighting its capability to effectively simulate the movement of the saturation front and the corresponding changes in pore pressure distribution under varying conditions.

2.3 Slope Model Test

A slope model experiment was conducted to determine suitable K_f values (Li et al. 2025). Measurement points were arranged identically in both the experimental and numerical models (Fig. 2.13). Tensiometers and piezometers recorded changes in

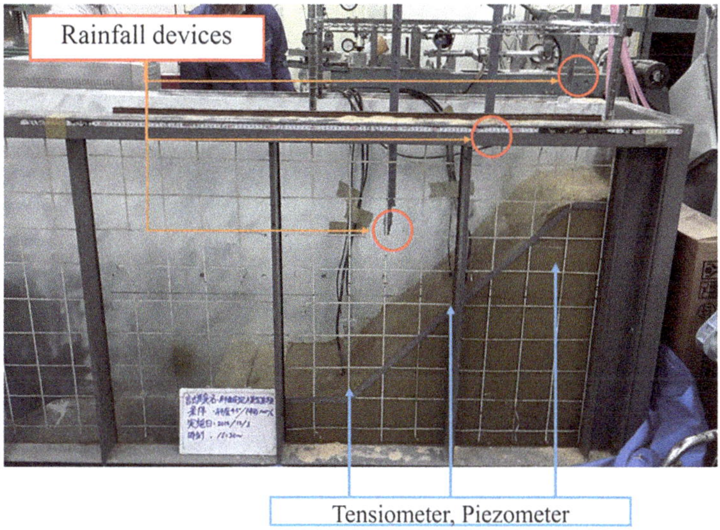

(a)

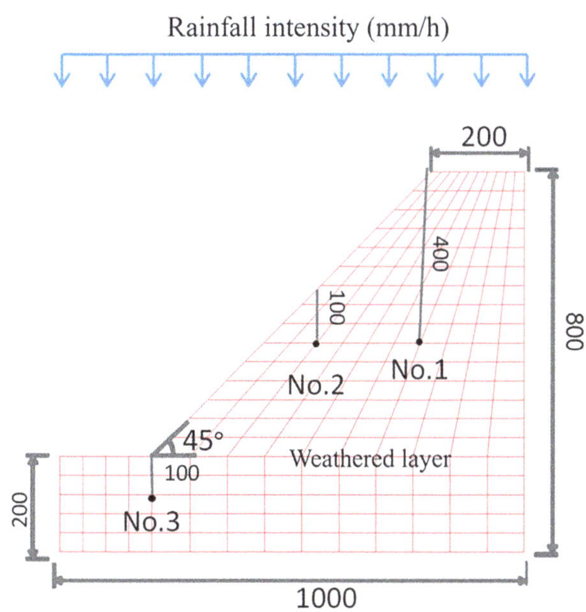

(b)

Fig. 2.13 (**a**) The slope model experiment under rainfall conditions. (**b**) The numerical model of slope and the positions of tensiometers and piezometers (unit: mm) (Li et al. 2025)

(a) (b)

Fig. 2.14 The nozzle for rainfall in the experiment with (**a**) 20 mm/h and (**b**) 40 mm/h

Table 2.2 The hydraulic boundary conditions of the slope model

Surfaces	Condition
Slope crest	Water supply
Slope surface	Water supply
Slope toe	Water supply
Bottom	No drainage
Right side	No drainage
Left side	Drainage

matric suction and pore water pressure. In single-phase flow analysis, negative pore pressure represents matric suction of soils. Variations in pore pressure, matric suction, and saturation were monitored under rainfall conditions. The initial moisture content of the remolded soil was set at 10% (Table 2.1). Boundary conditions included a permeable left boundary and impermeable right and base boundaries. This is consistent with the numerical model. The slope model was discretized into 384 zones, with rainfall applied to the slope surface at intensities of 20 mm/h and 40 mm/h (Figs. 2.13a and 2.14). Numerical results for saturation and pore pressure are shown in Fig. 2.13b. Table 2.2 summarizes the hydraulic boundary conditions for both the experiments and numerical simulations.

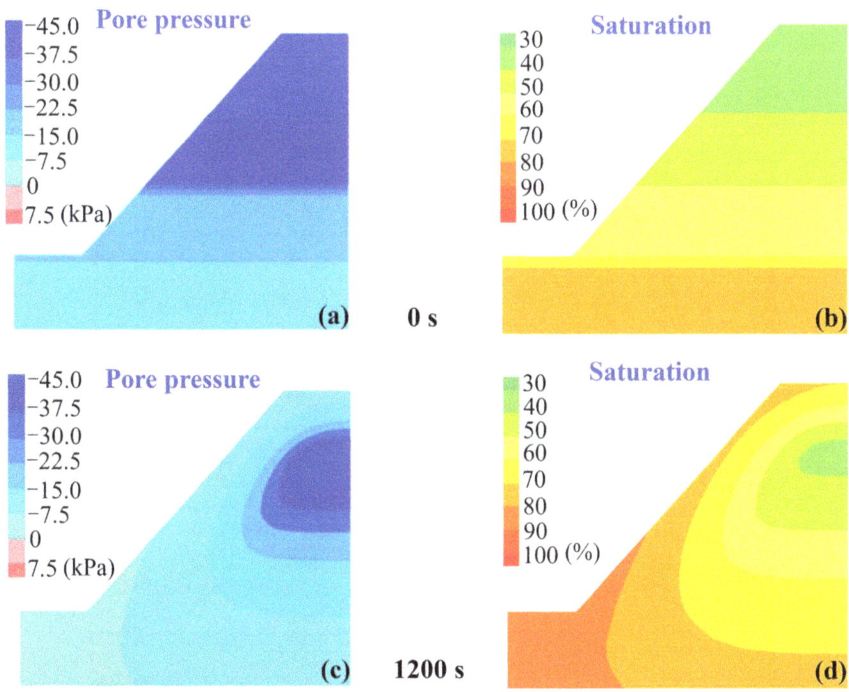

Fig. 2.15 The pore pressure (**a**) and saturation (**b**) of the model slope at initial stage and the pore pressure (**c**) and saturation (**d**) after 1200 s of rainfall (20 mm/h)

Comparisons between experimental and simulated results (Figs. 2.15 and 2.16) show close agreement. The fluid modulus K_f was set at 2×10^5 Pa based on these comparisons. During the early stages of rainfall, experimental pore pressures exhibited a delayed response compared with simulations. This reflects the time required for water to infiltrate to measurement points. Table 2.3 summarizes the differences between simulated and experimental values, indicating that the first dataset had the largest error and dispersion, while subsequent datasets showed smaller discrepancies. Minor deviations are attributed to the assumption of isotropic permeability in the model and the high permeability coefficient used in simulations. Overall, the close agreement confirms that the fitted VG parameters a and n reasonably represent the hydraulic properties of the weathered soil (Li et al. 2025).

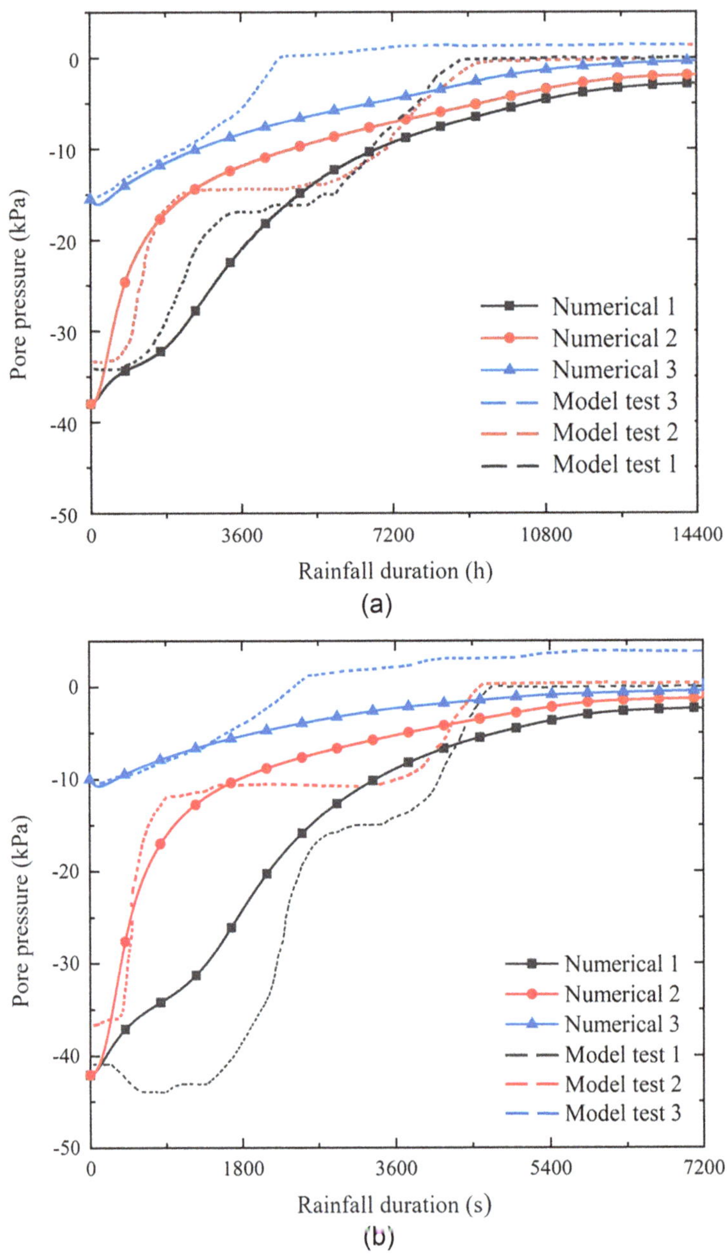

Fig. 2.16 Variation of pore pressure over time with rainfall intensities of (**a**) 20 mm/h and (**b**) 40 mm/h (Li et al. 2025)

Table 2.3 Quantifying errors between model tests and numerical simulation

Rainfall intensity	20 mm/h			40 mm/h		
Group	1	2	3	1	2	3
Mean error (kPa)	−15.3	−7.64	−5.22	−8.81	−9.70	2.67
Standard deviation	7.01	3.22	2.76	11.28	5.75	2.05

2.4 Conclusions

Through a series of laboratory experiments, this study focused on identifying three crucial hydraulic parameters that significantly influence the behavior of unsaturated soils under various conditions: the saturated permeability coefficient and the two fitting parameters, a and n, in the van Genuchten (VG) model. These parameters are essential for accurately modeling the hydraulic properties of unsaturated soils, particularly in the context of rainfall infiltration and slope stability analysis. The study demonstrated that the experimental results closely align with the numerical simulations, which validates the precision of these parameters. This alignment provides a solid foundation for subsequent numerical modeling, ensuring that the parameters used in the simulations are reliable and representative of actual field conditions. The importance of the saturated permeability coefficient lies in its role in governing the rate of water movement through the soil. This coefficient, when accurately determined, enables more precise predictions of infiltration rates and the subsequent distribution of water within the unsaturated zone. Similarly, the fitting parameters a and n in the VG model, which describe the soil's water retention curve and hydraulic conductivity, are crucial for simulating the behavior of water under unsaturated conditions. By accurately determining these parameters, the study ensures that the simulation results realistically capture the physical behavior of unsaturated soils, particularly under varying moisture conditions during rainfall infiltration.

To further validate the effectiveness of these parameters, a simple one-dimensional seepage analysis was conducted using FLAC^{3D}, a widely used commercial numerical modeling software. The results of this analysis confirmed the accuracy and reliability of FLAC^{3D} in solving unsaturated seepage problems, particularly with regard to modeling the free surface, which is a critical aspect in slope stability analysis. The ability of FLAC^{3D} to accurately capture the movement of the water table and the resulting pressure distribution is essential for assessing the stability of slopes under rainfall conditions. The comparison between the experimental results and the numerical simulations for a 2D slope model provided further support for the accuracy of the chosen parameters and the capabilities of FLAC^{3D}. The simulation results closely matched the experimental observations, confirming that the parameters derived from the laboratory tests were correctly implemented in the numerical model. This comparison highlighted the importance of using accurate input parameters in numerical simulations to ensure that the results are reflective of real-world conditions.

References

Cai F, Ugai K (2004) Numerical analysis of rainfall effects on slope stability. Int J Geomech 4:69–78. https://doi.org/10.1061/(ASCE)1532-3641(2004)4:2(69)

Kafle L, Xu W-J, Zeng S-Y, Nagel T (2022) A numerical investigation of slope stability influenced by the combined effects of reservoir water level fluctuations and precipitation: a case study of the bianjiazhai landslide in China. Eng Geol 297:106508. https://doi.org/10.1016/j.enggeo.2021.106508

Li X, Jiang Y, Sugimoto S (2025) Slope stability analysis under heavy rainfall considering the heterogeneity of weathered layers. Landslides 22:1257–1273. https://doi.org/10.1007/s10346-024-02404-8

Ng CWW, Shi Q (1998) A numerical investigation of the stability of unsaturated soil slopes subjected to transient seepage. Comput Geotech 22:1–28. https://doi.org/10.1016/S0266-352X(97)00036-0

Pan Y, Wu G, Zhao Z, He L (2020) Analysis of rock slope stability under rainfall conditions considering the water-induced weakening of rock. Comput Geotech 128:103806. https://doi.org/10.1016/j.compgeo.2020.103806

van Genuchten MT (1980) A closed-form equation for predicting the hydraulic conductivity of unsaturated soils. Soil Sci Soc Am J 44:892–898. https://doi.org/10.2136/sssaj1980.03615995004400050002x

Wei WB, Cheng YM, Li L (2009) Three-dimensional slope failure analysis by the strength reduction and limit equilibrium methods. Comput Geotech 36:70–80. https://doi.org/10.1016/j.compgeo.2008.03.003

Chapter 3
Theoretical Framework and Governing Equations

Abstract This chapter presents the theoretical framework and governing equations for slope stability analysis under rainfall conditions. It begins with a discussion of methods for simulating the heterogeneity of weathered soils, which is critical for accurately capturing spatial variations in slope material properties. The chapter then introduces the numerical implementation of unsaturated seepage analysis, including the treatment of matric suction based on Bishop's effective stress principle, as well as the constitutive models used to represent the mechanical behavior of the soils. Methods for representing the water-induced weakening effects of soft rock during rainfall are also described, highlighting their importance in simulating realistic responses of rock layers under rainfall conditions. Finally, case studies are presented in FLAC^{3D}, where the shear strength reduction method (SSRM) is employed to calculate safety factors for both two-dimensional (2D) and three-dimensional (3D) slopes under rainfall conditions. These examples illustrate the application of the theoretical framework and constitutive models within numerical simulations. They provide an accurate method of evaluating slope stability and yield valuable insights into the potential failure mechanisms under varying rainfall conditions.

Keywords Heterogeneity · Unsaturated seepage · Constitutive equations · Water-induced weakening · SSRM

3.1 Heterogeneity of Weathered Layers

The presence of heterogeneity poses challenges in analyzing the stability and failure mechanisms of weathered rock slopes under rainfall condition. Therefore, a comprehensive understanding of the heterogeneity within the weathered rock mass is essential for accurate slope stability analysis (Ietto et al. 2018). They analyzed the data and discussed the relationships between weathering degree, seismic velocity, rock mass quality, and landslide susceptibility. It concluded that the degree of

Y. Jiang, X. Li, *Stability Evaluation of Slope Under Heavy Rainfall Based on Geometric Features*, Springer Natural Hazards,
https://doi.org/10.1007/978-981-95-8656-1_3

Fig. 3.1 (**a**) Completely weathered rock with corestones of highly weathered rock (Classes IV–V), (**b**) Typical weathered profiles of granitoid rocks (Ietto et al. 2018)

weathering significantly influences both the occurrence and distribution of landslides in the studied area. Figure 3.1 shows granitoid rocks of the rock slope, exhibiting varying degrees of weathering. It is clear that incorporating the heterogeneity introduced by these different weathering levels into the numerical simulation would substantially enhance the accuracy and reliability of the computational results. Many researchers used statistical methods in the numerical models to explore the heterogeneity property of rocks. In the field of rock engineering, the Weibull distribution is extensively applied to analyze the heterogeneous of rocks. Given the statistical regularities of mechanical parameters rocks in the macroscopic level often adhering to the Weibull distribution, numerous scholars have approached the modeling of heterogeneous rock properties by assuming that parameters of microscopic rock units follow the Weibull distribution.

Hence, in this study, the heterogeneity of weathered layers is modeled using the Weibull distribution, which is expressed by Eq. (3.1):

$$\varphi = \frac{m}{\sigma_0}\left(\frac{\sigma}{\sigma_0}\right)^{m-1} \exp\left[-\left(\frac{\sigma}{\sigma_0}\right)^{m}\right] \tag{3.1}$$

where σ denotes a specific mechanical property, such as strength or elastic modulus, σ_0 is the scale parameter, and m is the shape parameter that defines the form of the

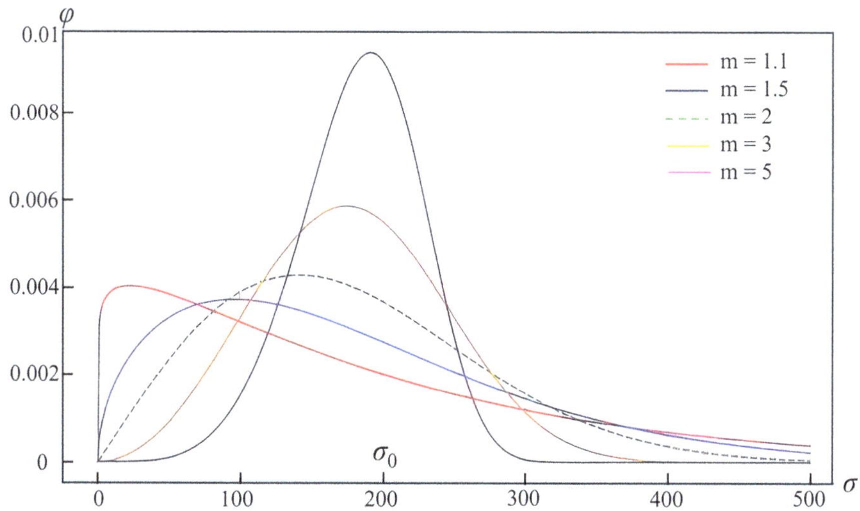

Fig. 3.2 The distribution of elastic modulus E (MPa) for five different heterogeneous specimens with different homogeneity index m

distribution. The parameter m, also referred to as the homogeneity index, indicates the degree of material uniformity. For instance, Fig. 3.2 illustrates the Weibull distribution of the elastic modulus E for rock samples with varying homogeneity indices (Tang et al. 2000). As m increases, the mechanical properties of the elements cluster more tightly around 200 MPa, indicating higher homogeneity. Conversely, a lower m value corresponds to greater variability in properties. Therefore, mechanical parameters such as cohesion c or elastic modulus E become more uniform with higher homogeneity indices and less uniform with lower indices (Li et al. 2025).

When $m = 1$, the Weibull distribution simplifies to an exponential distribution. This represents the highest level of heterogeneity or randomness in the material properties, meaning the mechanical parameters (e.g., shear strength, Young's modulus) are widely and randomly distributed. When $m = 3$, the distribution becomes more "peaked," meaning the material properties are less random and more homogeneous. As m increases, the variability decreases, and the parameters cluster closer to the mean value. For $m = 3$, the material exhibits a moderate level of heterogeneity, indicating some variability but not as extreme as $m = 1$. The Weibull distribution of mechanical properties in weathered layers is implemented through secondary development with FISH functions in FLAC^{3D}. This approach transforms a set of uniformly distributed random numbers in the interval [0, 1] into a corresponding set that follows the desired distribution. The procedure can be summarized as follows:

$$\sigma_i = \sigma_0 \left[\ln \frac{1}{1-\omega_i} \right]^{\frac{1}{m}} \tag{3.2}$$

where σ_i represents the strength or elastic modulus of element i, m is the shape parameter, and σ_0 is the scale parameter. A sequence of random numbers, ω_1, ω_2, …, ω_n is generated using the Monte Carlo method, with each value ranging between 0 and 1. n is the total number of elements.

Using the built-in random distribution function in FLAC^{3D}, the Weibull distribution of elastic modulus E and cohesion c is implemented through secondary development with FISH functions. Accordingly, the numerical model in this study assumes that the mechanical properties of the weathered rock layers follow a defined Weibull distribution. Although the internal friction angle φ of weathered soils also exhibits variability, its range is relatively narrow compared to that of E and c. Therefore, φ is treated as homogeneous within the weathered layer, consistent with previous studies (Nian et al. 2008; Chen et al. 2020). A linear distribution is applied to the shear strength parameters, with cohesion c increasing linearly with depth. The internal friction angle φ remains constant in the numerical simulation.

3.2 Unsaturated Seepage Analysis in FLAC^{3D}

In this study, a fluid–solid coupling model was adopted to simulate unsaturated seepage in the slope. Fluid–solid coupling is realized by activating the mechanical–fluid interaction module and adjusting the stiffness ratio R_k between the fluid modulus and the solid modulus (Itasca Consulting Group 2003). In unsaturated seepage analysis, establishing a relationship between saturation and matric suction is essential for determining the permeability in the unsaturated zone. Fluid transport is described by Darcy's law, which is widely used in numerical simulations of groundwater flow (Pan et al. 2020). The governing equation is expressed as follows:

$$q_i = -k_{il}\hat{k}(S)\left[p - \rho_f x_j g_j\right]_{,l} \tag{3.3}$$

where q_i is the specific discharge vector, $k_{il}\hat{k}(S)$ denotes the permeability tensor, with $\hat{k}(S)$ representing the relative mobility coefficient. p is the pore pressure, ρ_f is the fluid density, x_j is the Cartesian coordinate, and g_j is the gravity vector. The subscripts i, j, and l range from 1 to 3. The relative mobility coefficient $\hat{k}(S)$ is obtained from Eq. (3.4):

$$\hat{k}(S) = S^2(3 - 2S) \tag{3.4}$$

where S is the saturation.

Equation (3.4) is provided by the user manual of FLAC^{3D}. The influence of capillary pressure is neglected, and the relative fluid mobility is related to saturation by a cubic law. The permeability function which is based on Leong and Rahardjo (1997) provides an alternative method to calculate the unsaturated permeability $k(h)$:

$$k(h) = K_s S_e^{\beta} \tag{3.5}$$

where K_s is the saturated coefficient of permeability and β is the fitting parameter. S_e refers to the effective saturation of the soil. A comparative study of these two calculation methods will be conducted in future research.

The fluid mass balance equation is expressed as:

$$-q_{i,i} + q_v = \frac{\partial \zeta}{\partial t} \tag{3.6}$$

$q_{i,i}$ represents the flux intensity (1/s), q_v is the volumetric fluid source intensity (1/s), ζ denotes the change in water content, t represents time (s).

Since the thermal expansions was not considered in this study, the fully coupled fluid-solid analysis in the process of fluid calculation can be described by the constitutive equation which shows the relationship between pore pressure p, saturation S, volumetric strain ε, and changes in water content ζ can be expressed as:

$$\frac{1}{M}\frac{\partial p}{\partial t} + \frac{n}{S}\frac{\partial s}{\partial t} = \frac{1}{S}\frac{\partial \zeta}{\partial t} - \alpha\frac{\partial \varepsilon}{\partial t} \tag{3.7}$$

where M is the Biot modulus (Pa), n is the porosity, t is the time. When the compressibility of soil particles is not considered, the Biot coefficient α is set to 1.

It is crucial for fluid-solid coupling calculations to consider the changes in parameters such as cohesion, internal friction angle, and elastic modulus as the soil transitions from unsaturated to saturated conditions. This is because these parameters significantly influence the soil's mechanical behavior under varying moisture content, which in turn affects the stability and deformation characteristics of slopes. Firstly, as the wetting front advances into the slope, pore air migration occurs, leading to a decrease in matric suction in the unsaturated zone of the slope. This reduction in matric suction, combined with the increase in soil weight, lowers the effective stress and consequently reduces the soil's shear strength, which in turn decreases the stability of slopes. Secondly, soil deformation alters the microstructure of the soil medium, affecting water-air transport pathways, permeability characteristics, and seepage processes within the soil. Therefore, the fluid-solid coupling refers to the interaction between fluid and solid, where fluid flow affects the stress and deformation of the solid, and the deformation of the solid, in turn, influences fluid flow. In fluid-solid coupling analysis, the interplay between water infiltration and the mechanical property of the soil affects slope stability. In unsaturated conditions, the cohesion and frictional resistance between soil particles are influenced by capillary forces and the matric suction, which contribute to the overall strength of the soil. As the soil transitions to a saturated state, the capillary forces decrease and the soil's effective stress changes. This leads to a reduction in cohesion and friction. The internal friction angle and elastic modulus may also change due to the alteration of the soil structure and particle interactions as the pore spaces fill with water.

The equation for the VG model (van Genuchten 1980) shows the relationship between volumetric moisture content and negative pore pressure p in soils, which is given by:

$$\theta = \theta_r + \frac{\theta_S - \theta_r}{\left[1 + (p/a)^n\right]^m} \tag{3.8}$$

where θ denotes the volumetric water content. The other parameters involved in this relationship include θ_r (residual volumetric water content), θ_s (saturated volumetric water content), and fitting parameters a, m and n. Specifically, m is defined as 1–(1/n), where n is greater than 1. Utilizing the relationship between volumetric water content and saturation $\theta = nS$, the correlation between saturation S and negative pore water pressure p can be derived as follows:

$$S = S_r + \frac{1 - S_r}{\left[1 + (p/a)^n\right]^m} \tag{3.9}$$

where S_r is the residual saturation.

In FLAC^{3D}, the element stress is defined as positive in tension, and the pore water pressure is defined as positive in compression, consistent with the definition in the physical model proposed by (Zienkiewicz et al. 1997). The static equilibrium state is expressed using total stress:

$$\frac{\partial \sigma_{ij}}{\partial x_j} + \rho_s g_i = 0 \tag{3.10}$$

where ρ_s is the saturated density, which can be calculated based on the dry density of solid phase ρ_d and the fluid density ρ_w:

$$\rho_s = \rho_d + nS\rho_w \tag{3.11}$$

The Biot effective stress equation for saturated soil ($S = 1$) under single-phase flow is expressed as follows:

$$\sigma'_{ij} = \sigma_{ij} + \alpha p \delta_{ij} \tag{3.12}$$

where σ'_{ij} is effective stress, δ_{ij} is Kronecker symbol, and p is matric suction, defined as $p = u_a - u_w$, where u_a is the pore air pressure and u_w is the pore water pressure. The pore water pressure is positive under compression and negative under tension, which is opposite to stresses. The numerical simulation was conducted under single-phase conditions, which means the pore air pressure $u_a = 0$. When the compressibility of soil particles is not taken into account, the Biot coefficient α is set to 1.

Substituting Eqs. (3.11) and (3.12) into Eq. (3.10), the equilibrium equation for the soil structure in a saturated state is obtained:

$$\frac{\partial \sigma'_{ij}}{\partial x_j} + \rho_d g_i - \alpha(1-n)\frac{\partial p}{\partial x_i} - n\gamma_w \frac{\partial \phi}{\partial x_i} = 0 \tag{3.13}$$

where γ_w is the unit weight of the fluid, $\gamma_w = \rho_w g$, g is the gravitational acceleration magnitude, ϕ is the hydraulic head which is expressed as follows:

$$\phi = \frac{\alpha p}{\rho_w g} - \frac{x_k g_k}{g} \tag{3.14}$$

Considering the impact of negative pore pressure on effective stress is crucial for accurately assessing slope stability under rainfall conditions. According to the generalized effective stress definition (Schrefler and Scotta 2001), the effective stress of unsaturated soil can be expressed as follows:

$$\sigma'_{ij} = \sigma_{ij} + \alpha S p \delta_{ij} \tag{3.15}$$

In FLAC^{3D}, it is assumed that the saturation is equal to 1 at nodes where the pore water pressure is not equal to 0. From Eq. (3.11), it can be seen that FLAC^{3D} can automatically account for the density of elements in unsaturated zone by modifying the node's saturation using FISH functions. However, the modified saturation based on the secondary development of FISH functions can only be stored in extra variables for the unsaturated seepage analysis in FLAC^{3D}. The program still utilizes default values of 0 and 1 for saturation in the calculation of effective stress. Although the program sets the saturation of unsaturated soil to 1 in FLAC^{3D}, the FISH functions can be used to subtract the term $\alpha(1 - S)\delta_{ij}p$ from the total stress σ_{ij}, as illustrated in Eq. (3.16). It can be seen that the corrected Eq. (3.16) and Eq. (3.15) have the same form.

$$\sigma'_{ij} = \sigma_{ij} - \alpha(1-S)p\delta_{ij} + \alpha p \delta_{ij} = \sigma_{ij} + \alpha S p \delta_{ij} \tag{3.16}$$

where α is the Biot's coefficient and $\alpha = 1$ is usually adopted when the compressibility of soil particles is not considered. p is matric suction which is equal to the negative pore water pressure.

Substituting Eqs. (3.11) and (3.15) into Eq. (3.10), the equilibrium equation for the soil structure in the unsaturated state is shown as:

$$\frac{\partial \sigma'_{ij}}{\partial x_j} + \rho_d g_i - \alpha(s-ns)\frac{\partial p}{\partial x_i} - ns\gamma_w \frac{\partial \phi}{\partial x_i} = 0 \tag{3.17}$$

By comparing Eq. (3.13) with Eq. (3.17), it can be seen that when the saturation S approaches 1, Eq. (3.17) becomes the same as Eq. (3.13). This means that the

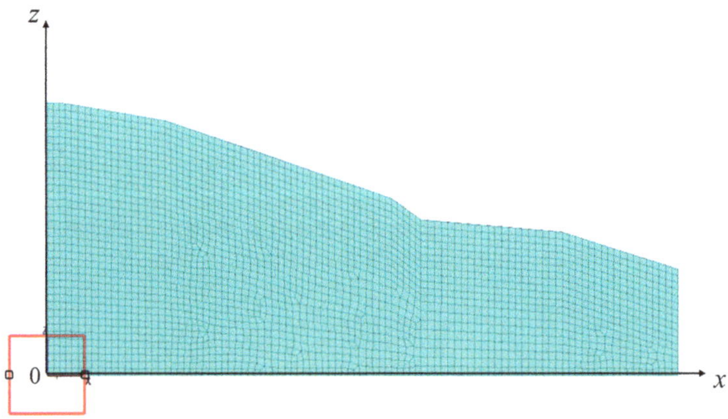

Fig. 3.3 The model is located within a coordinate system (x–z) in FLAC^{3D}

equilibrium equation in the unsaturated state becomes the equilibrium equation in the saturated state, which is consistent with the idea that "saturated soil can be regarded as a special case of unsaturated soil" Fredlund et al. (1978). The stress state of the soil significantly influences both its shear stiffness and shear strength for the unsaturated soils. Therefore, it is essential to correct the total stress to account for the impact of negative pore pressure on the soil's effective stress after performing saturated-unsaturated seepage calculations.

Thus, the total compressive stress in each element should be reduced by the product of α, $1 - S$, and p when the mechanical module computation is enabled in FLAC^{3D}. As shown in Fig. 3.3, if a two-dimensional slope model is considered, it is necessary to correct the effective stresses in the x and z directions within the xz coordinate system. The corrected total stress σ'_{xx} and σ'_{zz} is be expressed as ($\delta_{ij} = 1$):

$$\sigma'_{xx} = \sigma_{xx} - (1 - S)p \tag{3.18}$$

$$\sigma'_{zz} = \sigma_{zz} - (1 - S)p \tag{3.19}$$

3.3 Constitutive Equations

The explicit finite difference method was employed to simulate the coupled fluid–mechanical response of soil subjected to rainfall infiltration. The mechanical behavior was modeled using the Mohr–Coulomb constitutive framework, which in FLAC^{3D} is formulated in terms of the principal stresses σ_1, σ_2 and σ_3. The corresponding components of the generalized strain vector are the principal strains $\triangle_1$, $\triangle_2$ and $\triangle_3$. Based on these variables, the incremental form of Hooke's law can be expressed using the generalized stress and strain increments as follows (Pan et al. 2020):

$$\begin{cases} \Delta\sigma_1 = \alpha_1 \Delta e_1^e + \alpha_2\left(\Delta e_2^e + \Delta e_3^e\right) \\ \Delta\sigma_2 = \alpha_1 \Delta e_2^e + \alpha_2\left(\Delta e_1^e + \Delta e_3^e\right) \\ \Delta\sigma_3 = \alpha_1 \Delta e_3^e + \alpha_2\left(\Delta e_1^e + \Delta e_2^e\right) \end{cases} \tag{3.20}$$

where α_1 and α_2 are material constants expressed in terms of the shear modulus G, and bulk modulus K. The equations are given by:

$$\alpha_1 = K + \frac{4G}{3} \tag{3.21}$$

$$\alpha_2 = K - \frac{2G}{3} \tag{3.22}$$

The shear modulus G and bulk modulus K can be determined from Young's modulus E and Poisson's ratio μ as follows:

$$G = \frac{E}{2(1+\mu)} \tag{3.23}$$

$$K = \frac{E}{3(1-2\mu)} \tag{3.24}$$

According to the failure envelope $f(\sigma_1,\sigma_3) = 0$, the Mohr–Coulomb failure criterion can be expressed as:

$$\sigma_1 - \sigma_3 N_\varphi - 2c\sqrt{N_\varphi} = 0 \tag{3.25}$$

$$N_\varphi = \frac{1+\sin\varphi}{1-\sin\varphi} \tag{3.26}$$

An elastic–perfectly plastic model incorporating Bishop's effective stress concept is adopted. Therefore, the primary parameters governing the stress–strain behavior and failure characteristics of soils include Young's modulus E, Poisson's ratio μ, cohesion c and friction angle φ. In mechanical calculations of FLAC^{3D}, a persistent nonzero unbalanced force indicates the onset of failure and plastic flow within the model. As this study focuses solely on rainfall-induced shear failure, both the tensile strength and dilation angle of the soil are set to zero. This simplification ensures that no volumetric expansion occurs during shear deformation in the plastic phase.

The modified Mohr–Coulomb criterion based on Bishop's approach is used to capture changes in matric suction within the weathered soil during rainfall infiltration. The shear strength formulation for unsaturated soils proposed by Bishop (1959) is applied to compute the effective stress in the unsaturated zone, as expressed in Eq. (3.27):

$$\sigma' = (\sigma - u_a) + \chi (u_a - u_w) \tag{3.27}$$

where σ' is the effective normal stress, σ is the total normal stress, u_a is the pore air pressure, and u_w is the pore water pressure. The parameter χ ranges from 0 to 1 and varies with soil type and degree of saturation.

The shear strength can then be expressed as:

$$\tau_f = c' + \sigma' \tan\phi' \tag{3.28}$$

where τ_f is the shear strength of unsaturated soil, φ' is the effective internal friction angle, and c' is the effective cohesion. Because the shear strength of unsaturated soils is strongly influenced by the water content within soil pores (matric suction), Vanapalli et al. (1996) suggested that the χ factor can be reasonably approximated by the degree of saturation. Accordingly, the modified Mohr–Coulomb model based on Bishop's effective stress framework can be written as follows:

$$\tau_f = c' + (\sigma - u_a)_f \tan\phi' + S(u_a - u_w)_f \tan\phi' \tag{3.29}$$

where u_a is the pore air pressure, u_w is the pore water pressure, and S is the degree of saturation. In the numerical simulation, a single-phase analysis was performed ($u_a = 0$).

The effective cohesion c^* of an element, accounting for the influence of matric suction, is determined using Eq. (3.29):

$$c^* = c' + S(u_a - u_w)_f \tan\phi' \tag{3.30}$$

The influence of matric suction on soil stiffness, particularly regarding wetting-induced volume changes is illustrated by Hoyos et al. (2015). The soil's shear modulus G varies with matric suction and can be described using the best-fit power regression functions proposed in their study. Based on their findings, the optimal power regression equation for G is expressed as follows:

$$G = C(u_a - u_w)^H \tag{3.31}$$

where G represents the shear modulus of unsaturated soil, and $u_a - u_w$ denotes matric suction. The parameters C and H are obtained through curve fitting. Based on experimental results, with a confining pressure of 100 kPa in triaxial tests, C and H were determined to be 12.921 and 0.274, respectively. Since the confining pressure

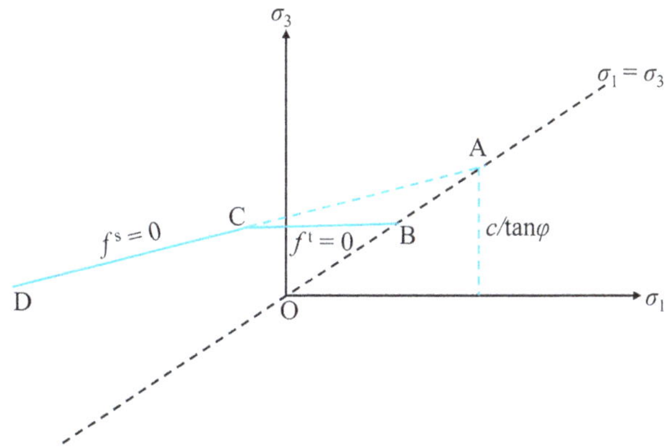

Fig. 3.4 Mohr-Coulomb failure criterion in FLAC^{3D} (Itasca Consulting Group 2003)

within the slope from the surface to a depth of approximately 5 m can be considered close to 100 kPa, these values are adopted in this study to model the effect of matric suction on soil stiffness.

The relationship between the elastic modulus E and shear modulus G is given by Eq. (3.23). Experimental data or literature regarding the variation of Poisson's ratio μ for unsaturated soils from the Sasebo landslide are currently lacking. Therefore, μ is assumed constant during rainfall infiltration. As shown in Fig. 3.4, the failure envelope in this model follows a Mohr-Coulomb criterion (shear yield function) with a tension cutoff (tension yield function) in FLAC^{3D}. Stress points on this envelope follow a non-associated flow rule for shear failure and an associated rule for tension failure. The original Mohr-Coulomb model can be extended to capture strain-softening behavior at large deformations (Robert 2017; Chen et al. 2021). A failure mechanism is indicated when a contiguous line of active plastic zones connects two surfaces, as shown in Fig. 3.5. The strain-softening and hardening behavior caused by soil-water interactions cannot be fully captured by the original criterion but can be addressed through a modified Mohr-Coulomb model. However, the application of this modified model in FLAC^{3D} is beyond the scope of this study and can be investigated in future research.

Here, we briefly introduce the constitutive law of the original Mohr-Coulomb (M-C) criterion in the plastic phase in FLAC^{3D}. The shear plastic flow is primarily represented through changes in the dilation angle. Since the dilation angle of the soil is simplified to 0 in our study, we only introduced the constitutive law for the elastic phase in the paper, as this study does not delve into the plastic phase of the soil. Clearly, the M-C criterion does not introduce net stress and suction as the two main stress variables, as the Barcelona Basic Mode (BBM) does. It also does not describe the complex mechanical response of unsaturated soils during the rainfall process by incorporating the yield, hardening, and plastic flow characteristics of the soil. Therefore, it is difficult to simulate the collapse upon wetting based on the original M-C criterion.

The flow rule is given a unique definition by application of the following technique. A line represented by the diagonal between the representation of $f^s = 0$ and f^t

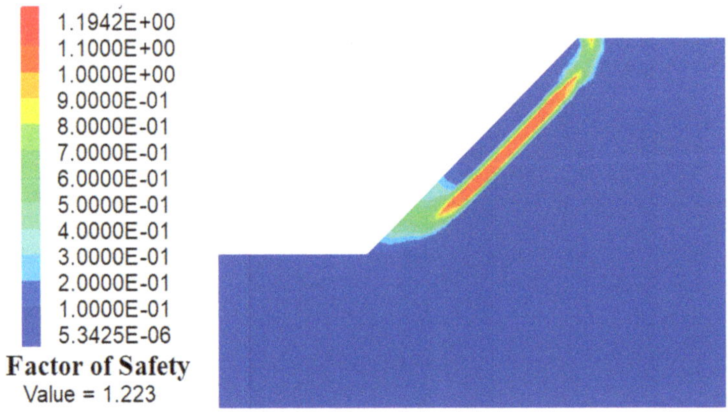

Fig. 3.5 The maximum shear strain of the slope after 48 h of rainfall

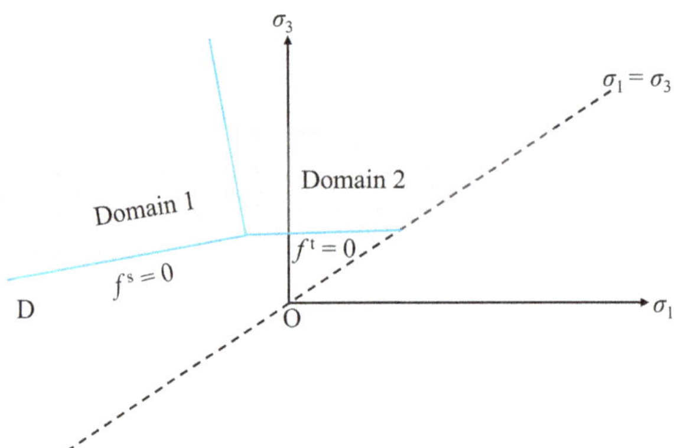

Fig. 3.6 Mohr-Coulomb model with domains used in the definition of the flow rule (Itasca Consulting Group 2003)

= 0 in the (σ_1,σ_3)-plane (see Fig. 3.6) divides the domain where an elastic guess violating the composite yield function into two domain: domain 1 and domain 2 (see Fig. 3.6). If the stress point falls within domain 1, shear failure is declared and the stress point is placed on the curve $f^s = 0$ using a flow rule derived using the potential function g^s. If the point falls within domain 2, tensile failure takes place, and the new stress point conforms to $f^t = 0$ using a flow rule derived using g^t. In this study, the tensile strength of the soil is set to 0, so the focus is solely on the plastic constitutive behavior related to shear, which corresponds to domain 1.

This study focuses primarily on shear failure under rainfall, which is the dominant mechanism in slope stability analysis. Consequently, tensile strength and dilation angle are simplified to focus on the shear behavior of slope. Limited experimental data for certain soil parameters, such as tensile strength and dilation angle, further justify this simplification. Moreover, the study addresses short-term rainfall-induced failure, where tensile effects and dilation are less significant than in long-term processes. The slope's factor of safety is evaluated using the shear strength reduction method (SSRM), with cohesion c and internal friction angle φ as the key parameters. Failure is indicated by a fully developed plastic shear band triggered by increased pore pressure, using the Mohr-Coulomb (M-C) criterion in FLAC^{3D} (Fig. 3.5). Once c and φ satisfy the yield criterion, the soil enters the plastic phase. Strain-softening and flow-like deformations associated with large strains are not considered in this study.

Wetting-induced collapse is significant in fine-grained soils such as loess or expansive silts (Zhou et al. 2014). It is less pronounced in the Sasebo landslide soil, which consists mainly of coarse-grained sub-angular gravel (10–30 mm) from sandstone. Nevertheless, future experiments based on the Barcelona Basic Model (BBM) are planned to investigate the influence of wetting-induced collapse on slope failure. Incorporating BBM in FLAC^{3D} would require developing a new constitutive model via FISH functions. For the current study, the Mohr-Coulomb criterion combined with Bishop's effective stress principle is employed to evaluate the influence of matric suction on slope stability under rainfall. This approach assumes that soil stiffness (elastic modulus E and shear modulus G) is primarily governed by effective stress and matric suction, rather than by volumetric changes associated with collapse (Wang et al. 2019). The ideal elastic-perfectly plastic model provides a simplified yet effective means to analyze short-term rainfall-induced failures in highly permeable soils, while more complex constitutive models such as BBM may be applied in future work to capture collapse and non-linear soil behavior.

3.4 Water-Induced Weakening of Rocks

Once rainwater infiltrates from the soil layer into the weathered rock layers of the slope, the theory of unsaturated soil mechanics is no longer applicable. Instead, the focus shifts to the water-induced weakening of soft rocks, which is based on the reduction of physical and mechanical parameters caused by the effects of water on soft rocks. Vasarhelyi (2003) analyzed the published data and showed that there is a linear correlation between the dry and fully saturated uniaxial compressive strengths, σ_{c0} and σ_{csat}, respectively. The overall best-fit equation for the 35 investigated sandstones is shown as:

$$\sigma_{\text{csat}} = 0.759\sigma_{\text{c0}}\left(R^2 = 0.906\right) \tag{3.43}$$

As shown in Fig. 3.7, there is a linear relationship between the uniaxial compressive strength (UCS) of standard rock samples and degree of saturation, meaning that the UCS decreases linearly as the saturation level increases (Li et al. 2012; Pan et al. 2020). For the convenience of calculation, the strength and Young's modulus of the wet rock specimens can be calculated with Eqs. (3.44) and (3.45), respectively:

$$\sigma_w(S) = \sigma_{Dry} - \left(\sigma_{Dry} - \sigma_{\text{Saturated}}\right)S \tag{3.44}$$

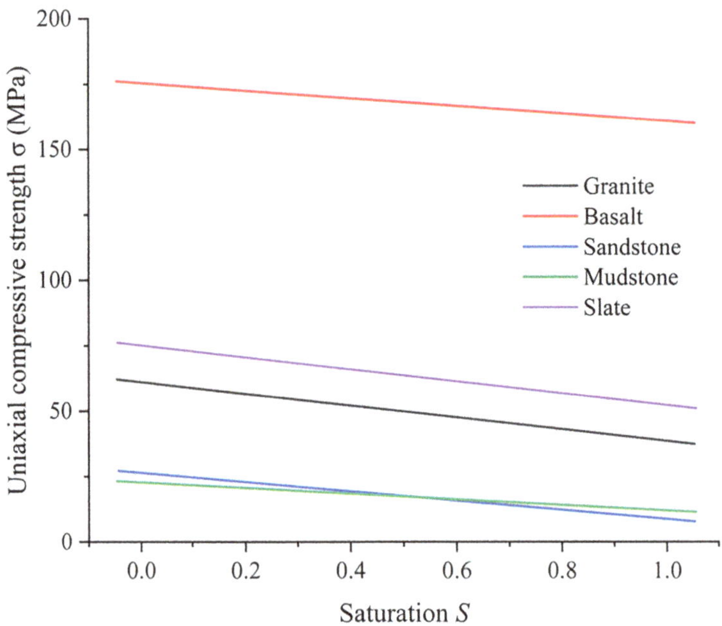

Fig. 3.7 Relationship between the uniaxial compressive strength and saturation (Pan et al. 2020)

$$E_w(S) = E_{Dry} - \left(E_{Dry} - E_{\text{Saturated}}\right)S \tag{3.45}$$

where $\sigma_w(S)$ is the UCS of rock material with a saturation of S, σ_{Dry} is the UCS of dry rock, $\sigma_{Saturated}$ is the UCS of saturated rock, $E_w(S)$ is the Young's modulus of rock material with a saturation of S, E_{Dry} is the Young's modulus of dry rock, and $E_{Saturated}$ is the Young's modulus of saturated rock.

The key parameters influencing the stress-strain relationship and failure behavior of rocks are Young's modulus, Poisson's ratio, cohesion, and the friction angle (Pan et al. 2020). Similar to strength and Young's modulus, the relationship between saturation and other mechanical parameters (Poisson's ratio μ, cohesion c, and friction angle φ) can be approximated as linear for ease of calculation (Fig. 3.8). The water-induced weakening effect can be quantified by a parameter λ, which represents the ratio of the rock's strength in a saturated state to its strength in a dry state. As illustrated in Fig. 3.9, an exponential relationship exists between the weakening coefficient λ and porosity of sandstones (Li et al. 2012; Yu et al. 2019; Eremin 2020; Pan et al. 2020). The weakening coefficient λ decreases to a certain extent and then gradually stabilizes as porosity decreases. It is worth noting that due to the difference in the degree of saturation between the samples of laboratory test and the actual slope during the rainfall process, the reduction factor λ for the same porosity in the numerical simulation is not as pronounced as shown in Fig. 3.9. Therefore, a more conservative reduction factor λ was used in the subsequent slope simulation calculations.

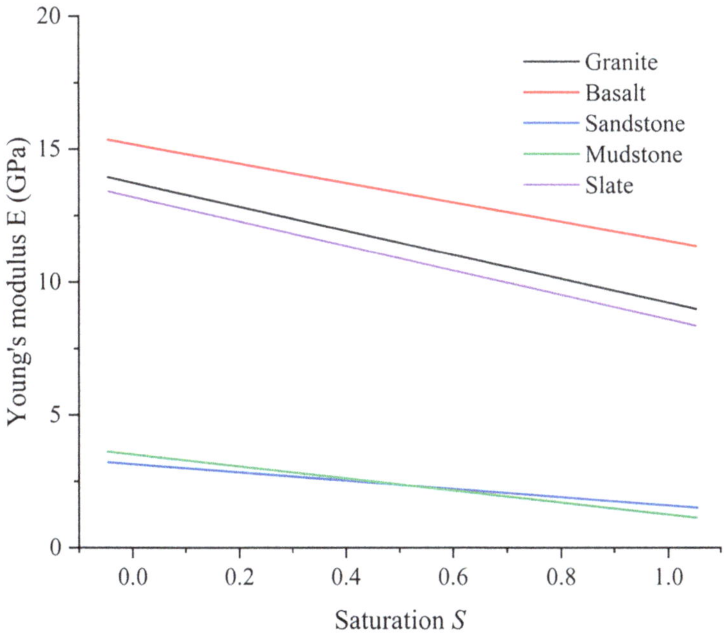

Fig. 3.8 Relationship between the Young's modulus and saturation (Pan et al. 2020)

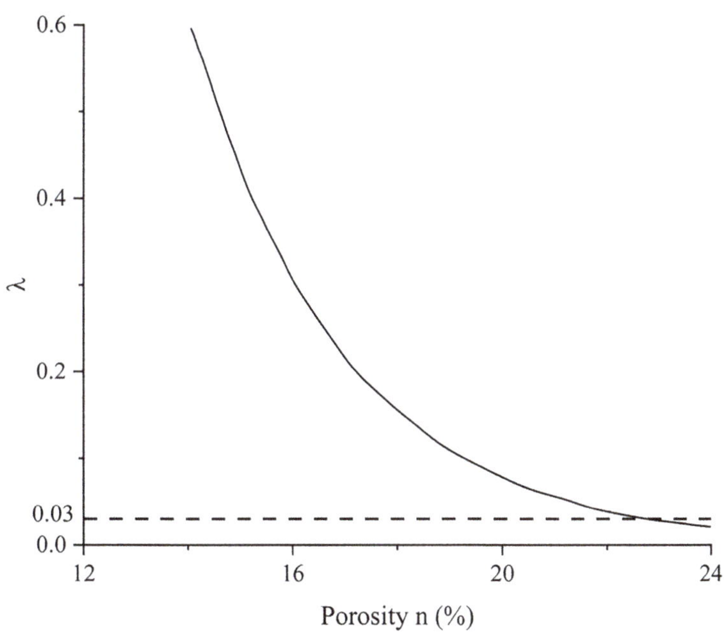

Fig. 3.9 Relationships between the porosity n and the weakening coefficient λ (Pan et al. 2020)

3.5 Slope Stability Analysis

Inside the soil mass forming the slope, the shear stress exerted by gravity is greater than that on a horizontal surface and acts in the direction that would cause the slope to fail. The distribution of shear stress within the slope is uneven, even in a state of equilibrium. A critical equilibrium condition typically develops at the crest of the slope, leading to localized failure. When localized failure occurs at the crest, it disrupts the equilibrium, causing stress concentration in adjacent areas. As a result, these areas eventually reach a critical condition, and the failure propagates downward. The localized failure at the crest then extends downward and outward, and the entire soil mass that forms the slope begins to slide along a curved surface or a set of surfaces. This curved surface, known as the slip surface, leads to the complete collapse of the slope. In addition to gravity, the forces acting to cause slope sliding typically include inertial forces during earthquakes, seepage forces from rainfall infiltration, and surface loads. These external forces all contribute to an increase in shear stress within the slope, potentially leading to collapse. On the other hand, excess pore water pressure that accumulates within the embankment during rainfall, or fluctuations in groundwater levels caused by rainfall can reduce the shear strength of soils. This also increases the likelihood of slope failure. Therefore, the problem of slope failure focuses on the balance between shear stress and the shear resistance of soils, while also evaluating the overall balance of forces acting on the entire slope, including external forces such as self-weight, groundwater pressure, and others. In slope stability analysis, the limit equilibrium method is widely used, and considerable practical results have been accumulated. Fellenius (1936) proposed a limit equilibrium analysis method for circular sliding, dividing the sliding soil mass into vertical slices and defining the safety factor using the following equation:

$$F_s = \frac{\tau_f}{\tau} \tag{3.46}$$

$$\tau_f = \Sigma\left\{c \cdot l + \left(W\cos\alpha - u \cdot l\right)\tan\phi\right\} \tag{3.47}$$

$$\tau = \Sigma\left(W\sin\alpha\right) \tag{3.48}$$

where F_s represents the factor of safety, τ_f denotes the shear strength, τ is the shear stress, c stands for the cohesion of the slip surface, l is the length of the slip surface of the slice, W refers to the weight of the slice, α is the angle of the slip surface, u indicates the pore water pressure, and ϕ represents the shear resistance angle of the slip surface, as shown in Fig. 3.10. It involves calculating the safety factor for a slope by dividing the resisting forces by the driving forces acting along a potential slip surface.

For fluid-soil coupled analysis in FLAC^{3D}, the determination of slope instability is based on the ratio of maximum unbalanced force, denoted as R_{unbal}. The unbalanced force is defined as the sum of forces acting on the nodes, which ideally

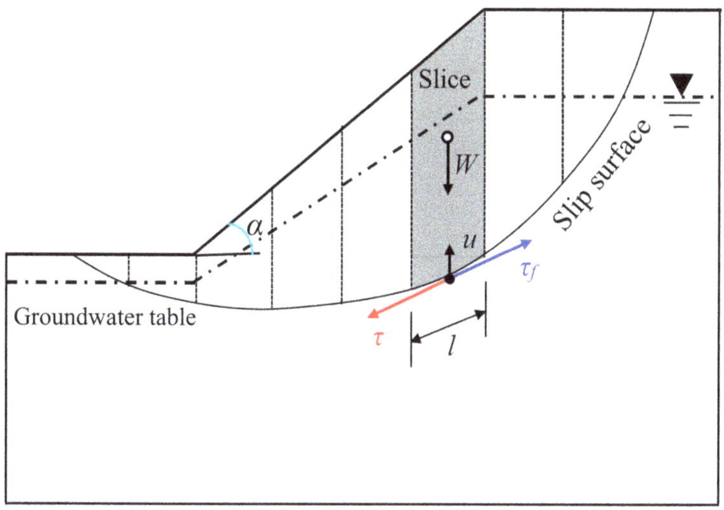

Fig. 3.10 Conceptual diagram of the limit equilibrium method (Fellenius Method)

approaches zero when the system is in equilibrium. The maximum unbalanced force ratio is defined as the ratio of the maximum unbalanced force in the entire model to the minimum internal force. A smaller value of this ratio indicates that forces are balanced across all nodes. Empirically, a maximum unbalanced force ratio $R_{\text{unbal}} \leq 1.0 \times 10^{-5}$ is considered a stable state, while values exceeding this threshold indicate localized instability. A slope model with a height of 40 m from the toe to the crest and a slope angle of 40° was used to demonstrate how the maximum unbalanced force ratio can be employed to judge slope instability (Fig. 3.11). In this model, the forces acting on the nodes are analyzed, and the unbalanced force, which represents the difference between the forces acting on the nodes and the equilibrium state, is calculated. As the rainfall progresses and the slope undergoes changes, the unbalanced force ratio increases. When the maximum unbalanced force ratio exceeds a critical threshold, it indicates that the slope has lost equilibrium and is transitioning toward instability. By monitoring this ratio, the onset of slope failure can be identified, providing a reliable method for assessing the stability of the slope under changing conditions. Rainfall with an intensity of 20 mm/h was applied to the surface of the slope. Figure 3.12 illustrates the time-dependent variation of the maximum unbalanced force ratio during the rainfall event. Approximately 72 h after the onset of rainfall, a rapid increase in the maximum unbalanced force was observed. This sharp rise indicated that the model could no longer maintain equilibrium, indicating the onset of slope instability. This behavior highlights the critical point at which the slope transitions from a stable to an unstable state under rainfall conditions.

The maximum shear strain distribution, shown by the red contours in Fig. 3.11, is used to estimate a failure surface. The slip surface is represented by the blue solid line, as shown in Fig. 3.11. Along this failure surface, the maximum principal stress difference is calculated for five elements labeled No. 1 to 5. The Mohr-Coulomb

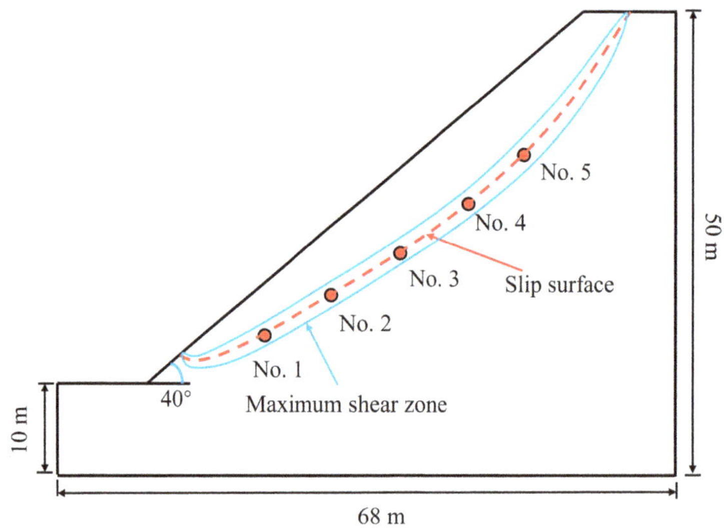

Fig. 3.11 Analysis model and maximum shear strain distribution after 72 h of rainfall

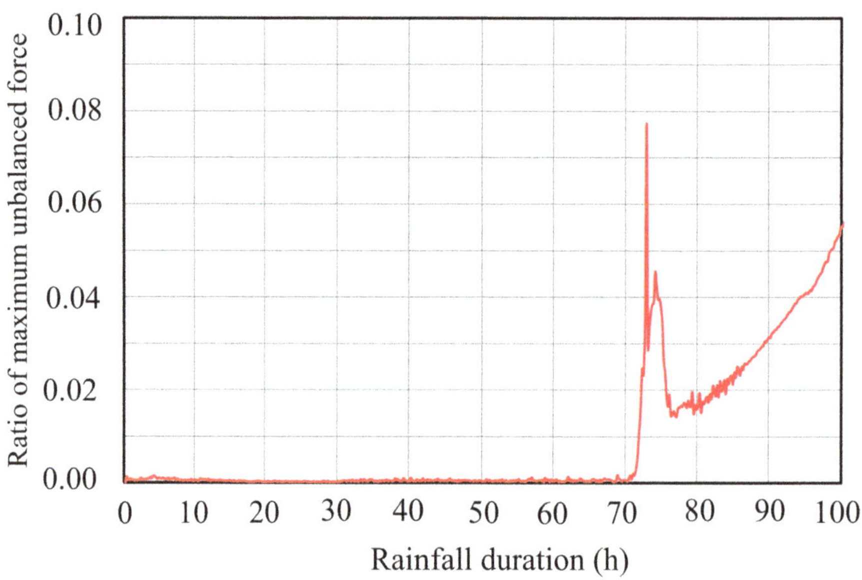

Fig. 3.12 Variation of maximum unbalanced force ratio with increasing rainfall duration (Ishida et al. 2018)

failure criterion is then applied to assess whether these elements have reached failure. Figure 3.13 shows Mohr's stress circles and the failure envelope at the time of failure. From this figure, it is confirmed that the relationships expressed in Eqs. (3.49) and (3.50) are valid. These equations accurately describe the stress

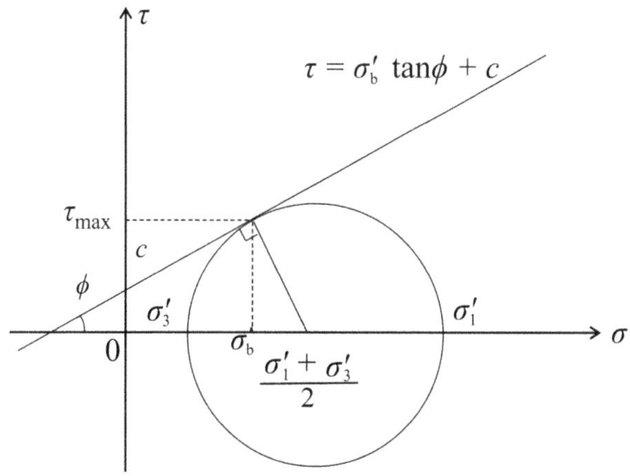

Fig. 3.13 The Mohr's stress circle at failure

conditions at the failure surface and align with the observed results, confirming the consistency of the failure analysis.

$$\sigma_b' = \frac{\sigma_1' + \sigma_3'}{2} - \frac{\sigma_1' - \sigma_3'}{2}\sin\phi \tag{3.49}$$

$$\tau_{\max} = \frac{\sigma_1' - \sigma_3'}{2}\cos\phi \tag{3.50}$$

where σ_1' and σ_3' represent the maximum and minimum principal stresses acting on the element, respectively. σ_b' is the normal stress, and ϕ represents the internal friction angle of soil. τ_{max} is the maximum shear stress. The failure envelope is expressed as:

$$\tau_{\max} = \sigma_b' \tan\phi + c \tag{3.51}$$

where c is the cohesion of soil.

By substituting Eqs. (3.49) and (3.50) into Eq. (3.51) and simplifying, the maximum principal stress difference at failure, $(\sigma_1' - \sigma_3')_{max}$, can be derived as follows:

$$(\sigma_1' - \sigma_3')_{\max} = (\sigma_1' + \sigma_3')\sin\phi + 2c\cos\phi \tag{3.52}$$

Based on the Eq. (3.52), Eq. (3.53) is valid when slope failure occurs.

$$(\sigma_1' + \sigma_3')\sin\phi + 2c\cos\phi = 0 \tag{3.53}$$

The calculation results of the left side of Eq. (3.53) for the five elements shown in Fig. 3.14, along with the variation of the maximum unbalanced force ratio over time are presented. As the calculation result of the left side of Eq. (3.53) approaches zero,

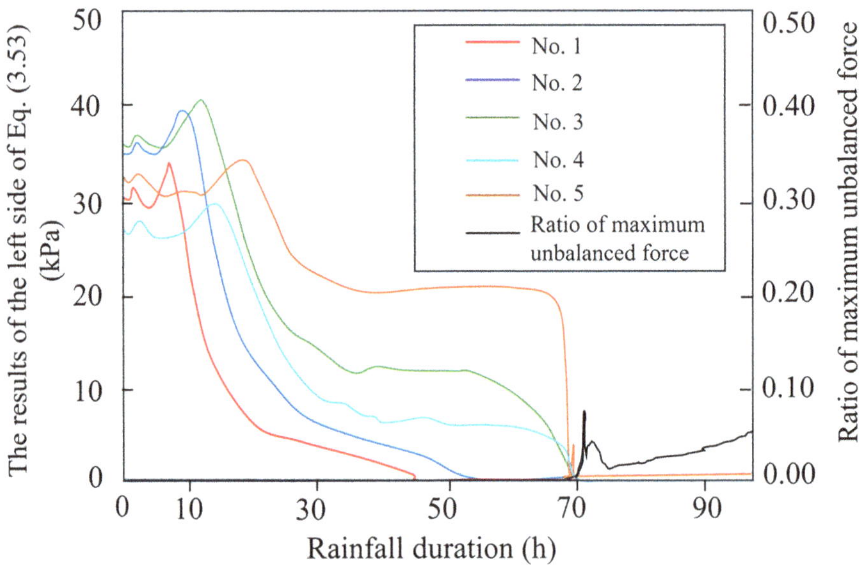

Fig. 3.14 The calculation results of the left-hand side of Eq. (3.53) and the variation of the maximum unbalanced force ratio with increasing rainfall duration

the corresponding element reaches a state of failure. Element No. 1, located at the foot of the slope where the maximum shear strain is most pronounced, is the first to fail. However, at this point, no overall slope instability has occurred. Approximately 72 h after the onset of rainfall, a sharp increase in the maximum unbalanced force ratio is observed, and three elements—No. 3, 4, and 5—fail almost simultaneously. This indicates that slope instability occurred at this time. Therefore, it can be concluded that the slope model became unstable when the maximum unbalanced force ratio increased sharply, as illustrated in Fig. 3.14.

3.5.1 Case Study (2D)

In this study, a single-phase flow analysis was conducted to reproduce the slope failure that occurred in July 2009 in Fukutsu Town, Fukuoka Prefecture, Japan. Figure 3.15 shows the longitudinal profile of the slope in the affected area. The debris at the lower part of the failed slope is observed to consist of two layers: old debris and newly collapsed material, suggesting that similar failure events have occurred in the past. The bedrock of the slope is greenstone, and the surface layer is weathered soil with a reddish-brown color. The source of the collapse at the head of the slope is the weathered part of the bedrock. It is assumed that the failure involved the old debris, which was carried along with the collapse.

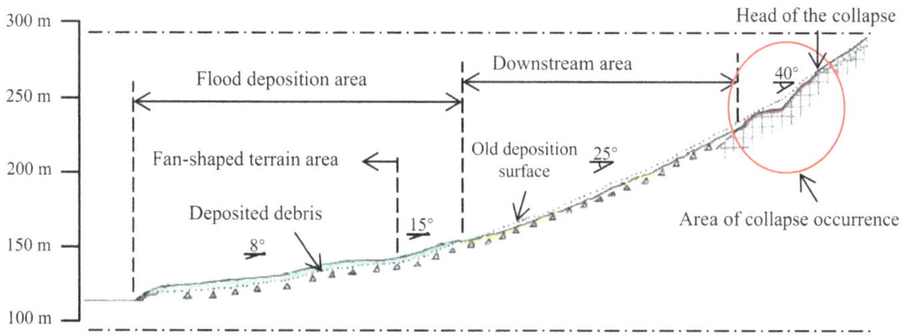

Fig. 3.15 Longitudinal section of the slope failure site in Fukuchi Town

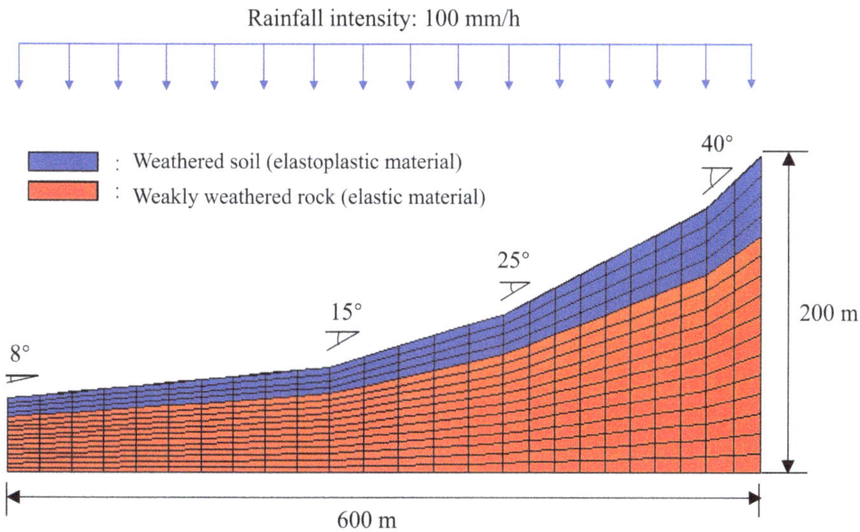

Fig. 3.16 Reconstructed analysis model of the slope failure in Fukuchi Town (Ishida et al. 2018)

Figure 3.16 shows the analysis model reproducing the site, and Table 3.1 presents the material properties used in the analysis. The material properties of the site are input based on the report by the Northern Kyushu Landslide Disaster Investigation Team of the Japan Geotechnical Society, and the analysis is conducted using these values. This report provides comprehensive data on the geotechnical characteristics of the region, including soil and rock types, shear strength parameters, and other relevant physical properties. These values are then used as inputs for the numerical analysis to simulate slope stability under various conditions, ensuring that the study reflects real-world conditions as closely as possible. The use of these values enhances the reliability and accuracy of the analysis.

Since the slope failure occurred in the weathered soil layer above the bedrock, the underlying slightly weathered rock is modeled as an elastic material, while the surface weathered soil is modeled as a fully elastic-plastic material according to the

Table 3.1 The mechanical parameters of slope in a saturated state

	Friction angle, °	Cohesion kPa	Poisson's ratio	Young's modulus MPa	Permeability m/s	Density (dry) kg/m^3	Porosity
Colluvial soil	36.9	12.7	0.30	200	2.66×10^{-5}	1800	0.3
Weakly weathered rock	38.0	100	0.13	300	5.0×10^{-7}	2300	0.15

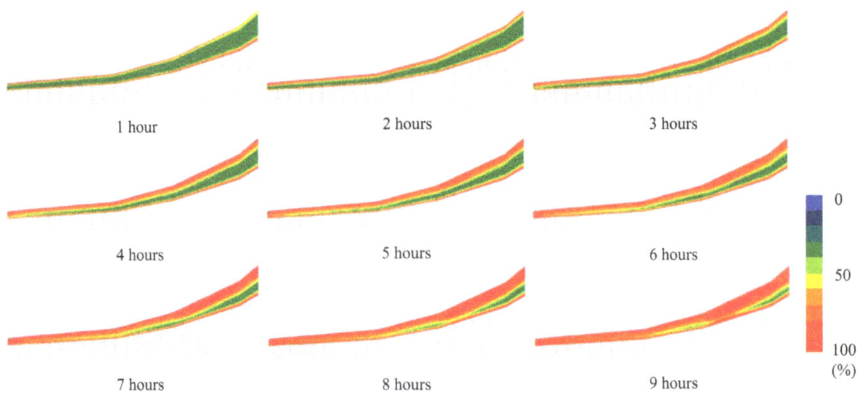

Fig. 3.17 Temporal changes in saturation distribution in the slope (Ishida et al. 2018)

Mohr-Coulomb failure criterion. For boundary conditions, the right side and the bottom of the model are set as non-draining, while the left side and the slope are draining. The bottom is restrained in both horizontal and vertical displacements, and the sides are restrained in horizontal displacement. The initial saturation of the slightly weathered rock and the weathered soil layers are set to 1.0 and 0.3, respectively. The rainfall intensity is set to 100 mm/h and input as the velocity head at the nodes from the slope surface. Additionally, due to the lack of experimental samples from the actual site, the parameters used in the simulation of 1D infiltration flow model in Sect. 2.2, are used here, with equivalent parameters from a similar sandy soil sample. Figure 3.17 shows the temporal changes in the saturation distribution up to 9 h after the rainfall begins. This figure shows how the saturation level of the soil changes over time. The infiltration front descends through the slope's surface as rainfall continues, and the overall saturation increases throughout the model. It can be observed that the infiltration front from the slope surface descends over time, and the overall saturation in the model increases. After 9 h of rainfall, the weathered soil layer becomes nearly fully saturated. Figure 3.18 presents the temporal changes in the pore water pressure distribution. Over time, the pore water pressure increases from the slope surface, with a particularly notable rise in pore water pressure near the failure zone, especially 3 h after rainfall begins. This figure illustrates the evolution of pore water pressure over time. As the rainfall continues, the pore water

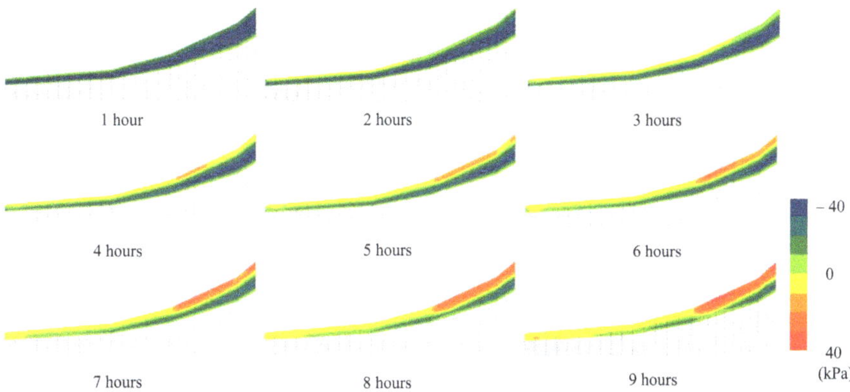

Fig. 3.18 Temporal changes in pore water pressure distribution in the slope (Ishida et al. 2018)

pressure gradually increases, especially near the point of slope failure initiation, where a significant rise is observed after 6 h of rainfall.

The causes of the landslide are considered from the following factors. Firstly, the increase in saturation due to rainfall. The infiltration from rainfall led to an increase in the saturation of the weathered soil layer, and by 9 h after the rainfall started, the entire weathered soil layer had nearly reached a saturated state. The rise in saturation increased the weight of the soil and elevated the sliding forces, which is considered to be a key trigger for the landslide. Furthermore, the infiltration caused an increase in pore water pressure at the surface of the slope, with a particularly significant rise near the landslide initiation area. These pore pressure changes reduced the shear resistance of the soil, compromising the instability of the slope and contributing to the landslide. Finally, the sudden increase in the maximum imbalance force ratio approximately 6 h after the rainfall began, a sharp increase in the maximum imbalance force ratio was observed. This suggests that the landslide occurred at this time. The increase in the imbalance force ratio indicates that the shear forces of the soil reached their limit, which led to the slope failure. The maximum shear strain increased in the shallow surface layer on the shoulder side of the slope. This indicates that the landslide was a surface failure. Field investigations also confirmed that the maximum depth of the landslide was around 3 m, which is consistent with the findings from the analysis. These combined factors contributed to the occurrence of the landslide. In particular, the changes in soil saturation and pore water pressure due to rainfall were key elements in the initiation of the landslide. As shown in Fig. 3.19, a sharp increase in maximum unbalanced force ratio is observed about 6 h after the start of the rainfall, suggesting that slope failure likely occurred around this time. Figure 3.20 shows the distribution of maximum shear strain approximately 6 h after the start of rainfall, corresponding to the time of failure. The maximum shear strain increases in the shallow layer of the slope, indicating that the failure mechanism is a surface failure. The maximum depth of the failure, about 3 m, is consistent with the observations by the Northern Kyushu Sediment Disaster

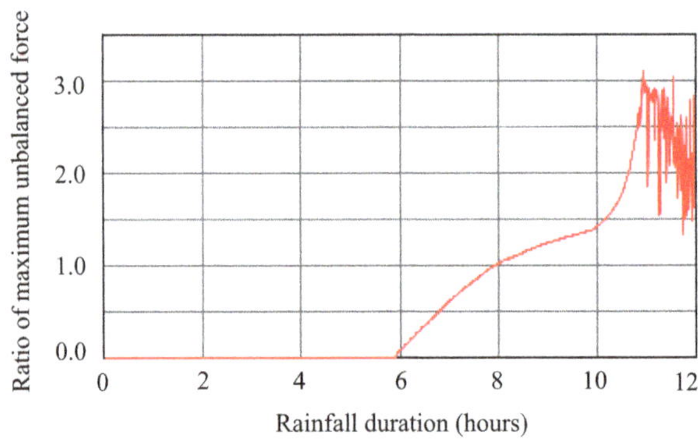

Fig. 3.19 Temporal changes in the maximum unbalanced force ratio

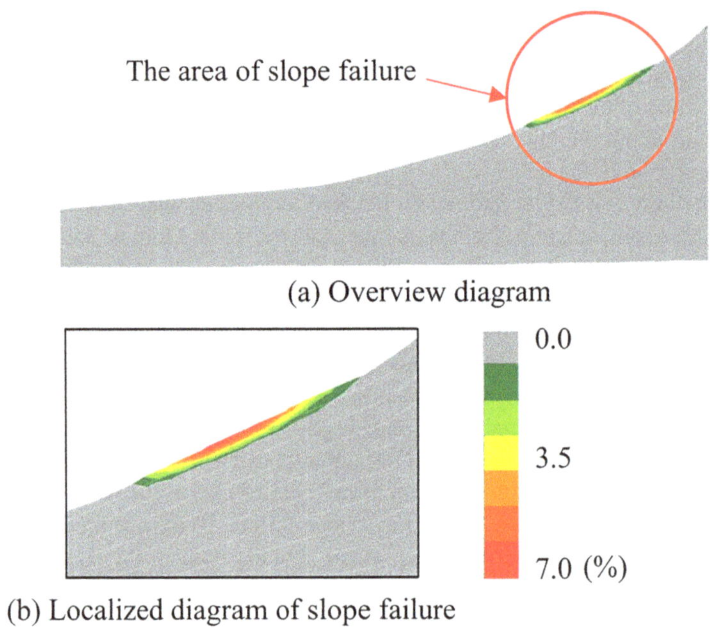

Fig. 3.20 Distribution of maximum shear strain after 6 h of rainfall (Ishida et al. 2018)

Investigation Team, which confirmed that the failure occurred in the surface layer. Additionally, comparing with the longitudinal slope section in Fig. 3.15, the location of the failure observed in the analysis aligns closely with the actual failure location. This demonstrates the correctness of selecting the physical and mechanical parameters of the slope materials, as well as the accuracy of the unsaturated rainfall infiltration simulation. This enhances the credibility of the analysis and affirms the effectiveness of the simulation approach in assessing slope stability.

Figure 3.18 shows that about 6 h after the start of rainfall, the increase in pore water pressure near the location where maximum shear strain occurs can be observed. This suggests that pore water pressures led to a reduction in effective stress, which subsequently decreased the shear strength and led to landslide occurrence. Based on the numerical analysis results, it can be observed that the landslide form and the location showed tendencies similar to those found in actual sites, indicating that it can represent the infiltration behavior of rainwater in situ. The landslide mechanism is caused by rainfall infiltration, which leads to an increase in pore water pressure near the landslide area, which causes a reduction in effective stress and thus leads to slope failure. However, since this analysis was carried out under a number of hypothetical conditions, future work should involve experimental measurements of pore water pressure for a more accurate representation of rainfall infiltration behavior in actual sites.

3.5.2 Shear Strength Reduction Method (SSRM)

The widely adopted method of shear strength reduction (SSR) was employed in the slope stability analysis (Zienkiewicz et al. 1975). The advantages of this method include its ability to model the formation of slip surfaces and capture local sliding behaviors. The shear strength reduction method is a technique that determines the overall safety factor of the slope through a total stress analysis. The overall safety factor, F_S, is an index that indicates the margin of safety of the slope before failure occurs. As shown in Fig. 3.21, F_S is defined by the ratio of the total sliding force acting on the sliding mass to the total maximum shear resistance on the slip surface, as expressed in Eq. (3.54).

$$F_s = \frac{\sum \text{the maximum shear resistance on the slip surface}}{\sum \text{the sliding force acting on the sliding soil mass}} \tag{3.54}$$

where F_S represents the overall safety factor of the slope.

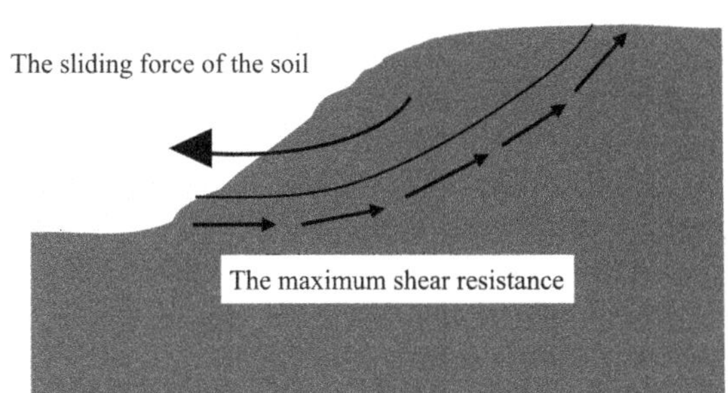

Fig. 3.21 The schematic diagram of force equilibrium in the slope

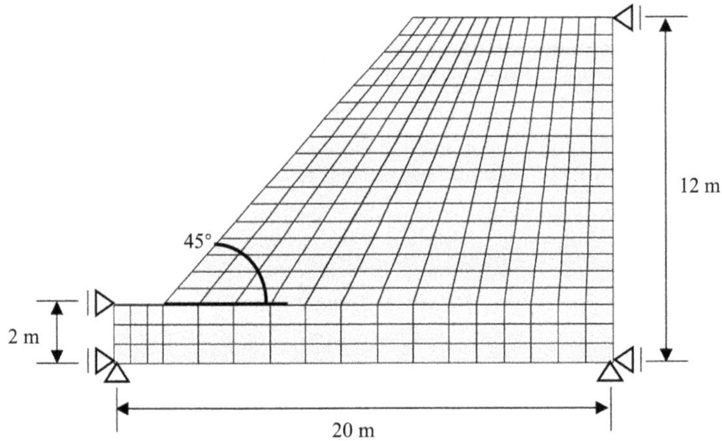

Fig. 3.22 The slope model built in FLAC^{3D}

When F_S is equal to 1, the slope reaches the limit state. Additionally, the larger the overall safety factor is compared to 1, the higher the safety of the slope against failure. This study primarily relies on variations in the factor of safety (FoS) to assess changes in slope stability. The expression for the shear strength reduction method is presented as follows:

$$\tau_m = \frac{c}{f} + \sigma_n' \times \frac{\tan\varphi}{f} \tag{3.55}$$

where σ_n' is the effective normal stress, c is the cohesion, and φ is the angle of internal friction. The value of f is adjusted until the slope fails, where the ultimate f is the factor of safety of the slope.

The safety factor (FoS) is calculated using the shear strength reduction method with a uniform slope model, as depicted in Fig. 3.22. The model has a height of 10 m and a slope angle of 45°. The material properties used in the analysis are shown in Table 3.2. The results show that when $F_S = 1.25$ (cohesion: 8 kPa, friction angle: 25°), the calculation could no longer converge in FLAC^{3D}. Thus, the overall safety factor (F_S) of the slope is evaluated to be 1.25. Figure 3.23 illustrates the distribution of maximum shear strain at the end of the calculation (after 10,000 cycles). Since failure is considered to occur at locations where the maximum shear strain increases, the critical slip surface is assumed to be the dashed line shown in the figure. By using the shear strength reduction method, the overall safety factor of the slope can be evaluated. The critical slip surface can be determined from the distribution of maximum shear strain at the end of the calculation, providing insight into the potential failure zones of the slope.

Table 3.2 The properties of slope material

Elastic modulus (MPa)	Poisson's ratio	Density (kg/m³)	Cohesion (kPa)	Internal friction angle (°)
200	0.3	2000	10	30

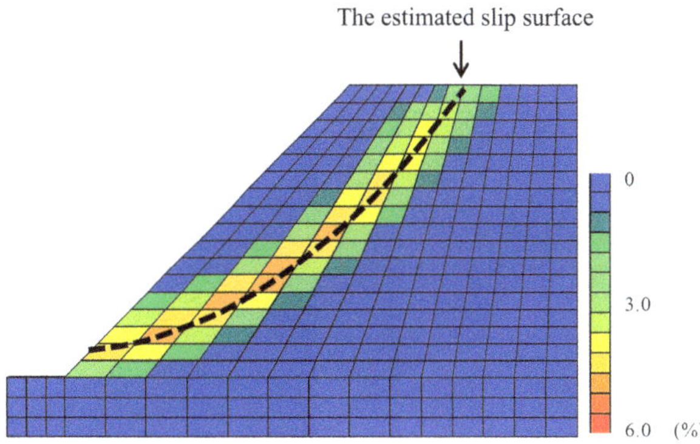

Fig. 3.23 The distribution of maximum shear strain at the end of the calculation

3.5.3 Case Study (3D)

As shown in Fig. 3.24, a three-dimensional slope model was developed to demonstrate the effectiveness of the shear strength reduction method. The rocky slope is situated along National Route 251, connecting Unzen City and Minamishimabara City in Nagasaki, Japan. The physical properties used in the analysis are listed in Table 3.3. These physical property values were obtained from field tests in previous studies and were uniformly applied to the entire model to identify hazardous areas based on terrain conditions. A constant rainfall intensity of 20 mm/h was applied to the slope surface in the numerical simulation. Matric suction in unsaturated soil slopes can be modeled using different assumptions. As shown in Fig. 3.25, Assumption A considers a linear increase in matric suction, starting from zero at the groundwater table and reaching a maximum value at the slope surface (Zhang et al. 2004). Assumption B, proposed by Xie et al. (2004), assumes that matric suction remains constant throughout the entire unsaturated zone. Assumption C, as suggested by Leong and Rahardjo (2012), posits that a constant matric suction above the maximum value provides a more accurate simulation of rainfall infiltration into a slope. Figures 3.26 and 3.27 illustrate the initial pore water pressure distributions corresponding to Assumptions A and B, respectively. For slopes with deep groundwater tables, a commonly adopted calculation model assumes that matric suction increases linearly to a certain depth above the water table, reaches a maximum value, and then remains constant. This study adopts Assumption B, setting the constant matric suction at −30 kPa based on the findings of Leong and Rahardjo (2012). Figure 3.28 illustrates the distribution of saturation in the slope before rainfall and after 12 h of rainfall. It can be observed that as rainfall

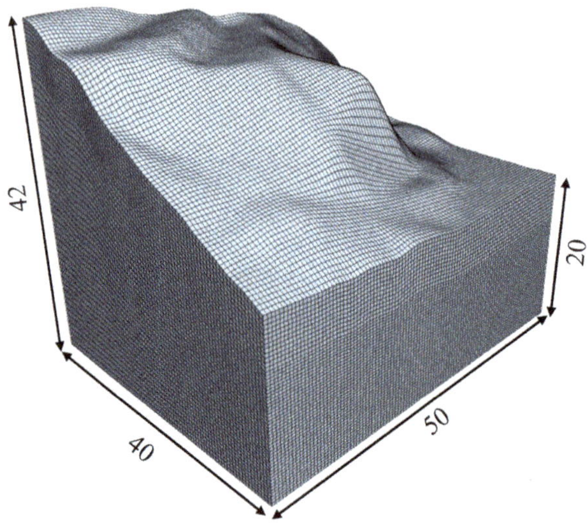

Fig. 3.24 Overview of the analysis model (unit: m)

Table 3.3 The mechanical parameters of slope in a saturated state

Weathered layers	Friction angle °	Cohesion kPa	Poisson's ratio	Young's modulus MPa	Permeability m/s	Density (Dry) kg/m³
Strongly	35	20	0.30	200	6×10^{-6}	1800
Moderately	38	30	0.20	300	1×10^{-6}	2100
Bedrock	40	100	0.13	600	5×10^{-7}	2200

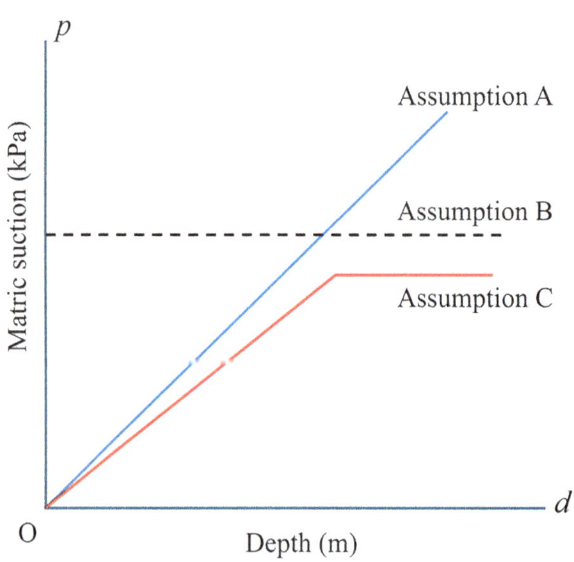

Fig. 3.25 Assumptions on distribution of matric suction in a slope

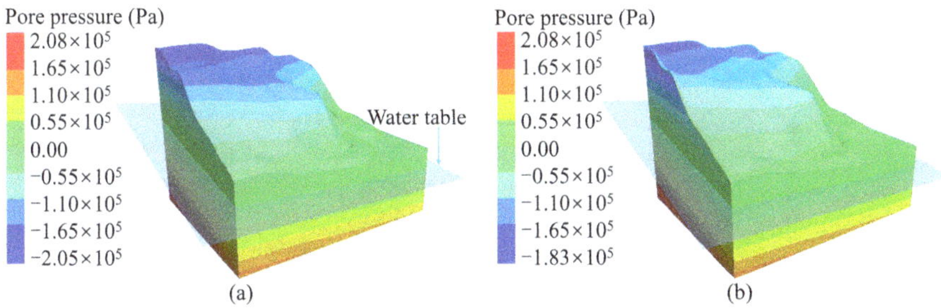

Fig. 3.26 Assumption A: distribution of pore pressure in the slope: (**a**) prior to rainfall and (**b**) after 12 h of rainfall

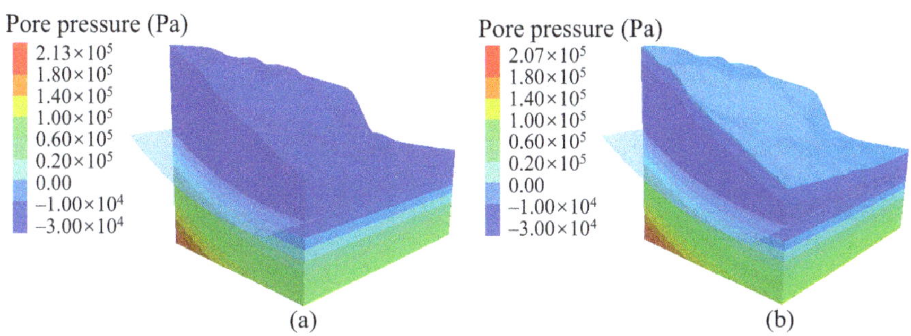

Fig. 3.27 Assumption B: distribution of pore pressure in the slope: (**a**) prior to rainfall and (**b**) after 12 h of rainfall

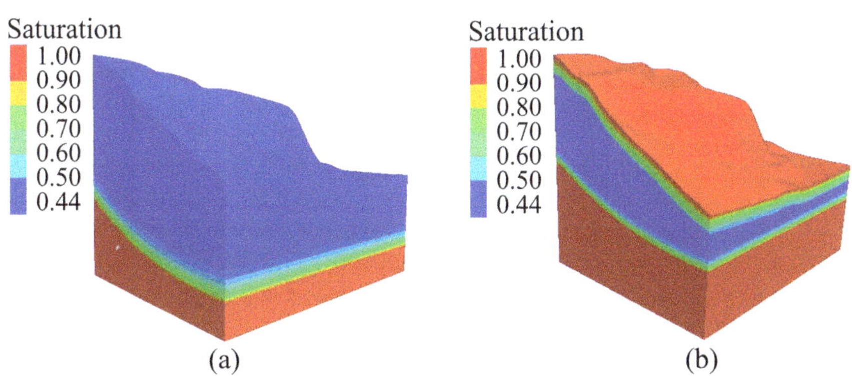

Fig. 3.28 Distribution of saturation in the slope: (**a**) prior to rainfall and (**b**) after 12 h of rainfall

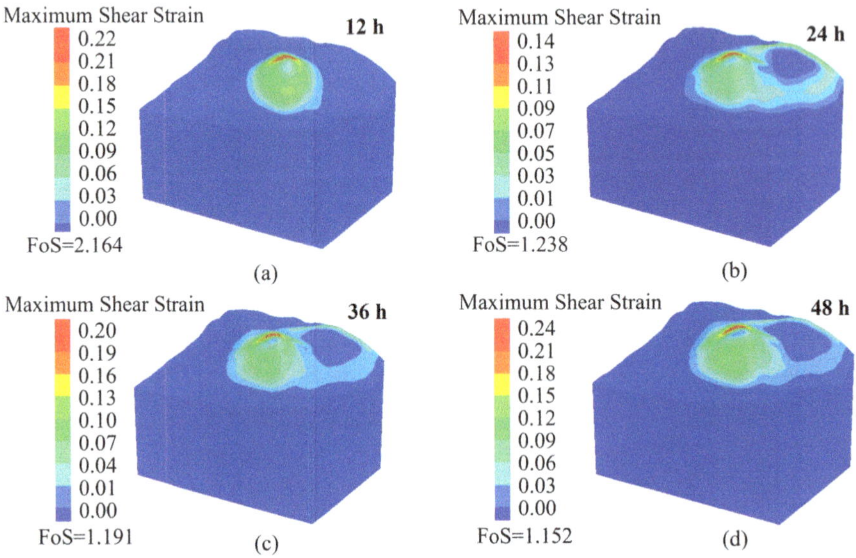

Fig. 3.29 Maximum shear strain and safety factor of the slope after 12 h (**a**), 24 h (**b**), 36 h (**c**) and 48 h (**d**) of rainfall

progresses, the unsaturated zone above the groundwater table within the slope gradually diminishes and transitions into a saturated zone. Figure 3.29 shows the safety factor and maximum shear strain contours obtained using the strength reduction method after 12, 24, 36, and 48 h of rainfall. It is clear that the safety factor of the slope gradually decreases, and the region of maximum shear strain expands with the increase in rainfall duration. This illustrates the significant impact of rainfall on the stability of the slope. Figure 3.30 shows the changes in the safety factor of the three-dimensional slope with the increase in rainfall duration. A significant decrease in the safety factor is observed between 12 and 24 h of rainfall. This is because the pore water pressure in the shallow layers of the slope increases rapidly within 24 h of rainfall, leading to a reduction in effective stress and a decrease in shear strength. After 24 h of rainfall, the change in the safety factor becomes less significant, as the large difference in permeability coefficients between the weathered layer and the bedrock slows the infiltration of rainwater into the bedrock. As a result, the pore water pressure in the bedrock does not change significantly and the change in the safety factor of the slope becomes less significant. This conclusion is consistent with the findings of Li et al. (2024). This also demonstrates the significant impact of the permeability coefficient of the soil and rock mass on slope stability under rainfall conditions.

Fig. 3.30 Variation of safety factor of the slope after rainfall

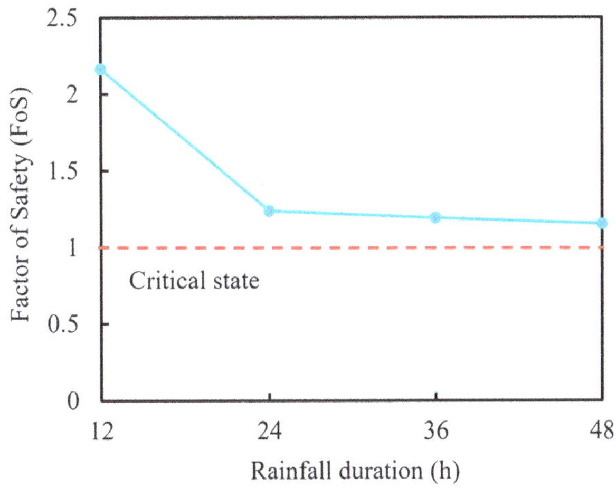

3.6 Conclusions

This chapter begins by introducing the application of built-in random distribution in FLAC^{3D} to implement the Weibull distribution of weathered layers in slope models. This approach enables numerical simulations to better reflect the heterogeneous characteristics of real-world slopes. A detailed methodology for simulating unsaturated seepage in FLAC^{3D} is then presented, highlighting the dynamic adjustment of permeability and saturation in the unsaturated zone throughout the rainfall process. The implementation also incorporates dynamic rainfall boundary conditions, ensuring that the slope surface remains under low pore pressure during the simulation. To analyze unsaturated seepage in the weathered soil layers of slopes under rainfall conditions, the modified Mohr-Coulomb criterion, based on Bishop's effective stress principle, is employed. Additionally, FISH functions are utilized to perform effective stress corrections under fluid-solid coupling conditions, significantly enhancing the accuracy and realism of the simulations. An elastic-perfectly plastic model based on Bishop's effective stress approach is used in this study. The explicit finite difference method is adopted to analyze the fluid-mechanical interactions in soil under rainfall effects, with the Mohr-Coulomb model serving as the failure criterion. However, once rainwater infiltrates from the weathered soil layer into the underlying rock layer, the unsaturated soil calculation theory becomes less applicable. This chapter further explores stability changes in rock slopes under rainfall infiltration, focusing on the reduction of physical and mechanical parameters caused by water-induced weakening effects in soft rocks. Variations in the porosity of weathered layers are considered through different reduction factors, λ, as rocks with higher porosity generally exhibit more pronounced weakening effects due to water saturation. Building upon these theoretical frameworks, the chapter presents two case studies—a 2D and a 3D slope stability analysis under rainfall conditions—using the strength reduction method to evaluate changes in slope safety factors during the rainfall infiltration process. The results of the 2D simulation reveal that the region of maximum shear strain (potential sliding surface) closely aligns with actual failure locations observed in real-world slopes, validating the reliability of the calculations. Moreover, the approach is extended to 3D slope stability analysis under rainfall conditions. The strength reduction method proves effective in identifying potential sliding regions and capturing the variations in safety factors as rainfall progresses. This comprehensive analysis offers valuable insights into the behavior of slopes under rainfall conditions, providing a robust framework for understanding the mechanism of rainfall-induced slope instability.

References

Bishop A (1959) The principle of effective stress. Teknisk Ukeblad 106:859–863

Chen G-H, Zou J-F, Pan Q-J et al (2020) Earthquake-induced slope displacements in heterogeneous soils with tensile strength cut-off. Comput Geotech 124:103637. https://doi.org/10.1016/j.compgeo.2020.103637

Chen X, Zhang L, Zhang L et al (2021) Modelling rainfall-induced landslides from initiation of instability to post-failure. Comput Geotech 129:103877. https://doi.org/10.1016/j.compgeo.2020.103877

Eremin M (2020) Three-dimensional finite-difference analysis of deformation and failure of weak porous sandstones subjected to uniaxial compression. Int J Rock Mech Min Sci 133:104412. https://doi.org/10.1016/j.ijrmms.2020.104412

Fellenius W (1936) Calculation of stability of earth dam. In: Proceedings of the Second Congress of Large Dams, pp 445–463

Fredlund DG, Morgenstern NR, Widger RA (1978) The shear strength of unsaturated soils. Can Geotech J 15:313–321. https://doi.org/10.1139/t78-029

Hoyos LR, Suescún-Florez EA, Puppala AJ (2015) Stiffness of intermediate unsaturated soil from simultaneous suction-controlled resonant column and bender element testing. Eng Geol 188:10–28. https://doi.org/10.1016/j.enggeo.2015.01.014

Ietto F, Perri F, Cella F (2018) Weathering characterization for landslides modeling in granitoid rock masses of the capo Vaticano promontory (Calabria, Italy). Landslides 15:43–62. https://doi.org/10.1007/s10346-017-0860-5

Ishida J, Jiang Y, Sugimoto S (2018) Basic study on use of two-phase flow method in slope stability analysis. J Japan Soc Eng Geol 59:263–272

Itasca Consulting Group I (2003) Online manual of FLAC3D fast Lagrangian analysis of continua in 3 dimensions. Itasca Consulting Group I, Minneapolis, MN

Leong EC, Rahardjo H (1997) Review of soil-water characteristic curve equations. J Geotech Geoenviron Eng 123:1106–1117. https://doi.org/10.1061/(ASCE)1090-0241(1997)123:12(1106)

Leong EC, Rahardjo H (2012) Two and three-dimensional slope stability reanalyses of Bukit Batok slope. Computers and Geotechnics 42:81–88. https://doi.org/10.1016/j.compgeo.2012.01.001

Li D, Wong LNY, Liu G, Zhang X (2012) Influence of water content and anisotropy on the strength and deformability of low porosity meta-sedimentary rocks under triaxial compression. Eng Geol 126:46–66. https://doi.org/10.1016/j.enggeo.2011.12.009

Li X, Jiang Y, Sugimoto S (2024) Slope stability analysis under heavy rainfall considering the heterogeneity of weathered layers. Landslides 22:1257–1273. https://doi.org/10.1007/s10346-024-02404-8

Li X, Jiang Y, Sugimoto S (2025) Slope stability analysis under heavy rainfall considering the heterogeneity of weathered layers. Landslides 22:1257–1273. https://doi.org/10.1007/s10346-024-02404-8

Nian TK, Chen GQ, Luan MT et al (2008) Limit analysis of the stability of slopes reinforced with piles against landslide in nonhomogeneous and anisotropic soils. Can Geotech J 45:1092–1103. https://doi.org/10.1139/T08-042

Pan Y, Wu G, Zhao Z, He L (2020) Analysis of rock slope stability under rainfall conditions considering the water-induced weakening of rock. Comput Geotech 128:103806. https://doi.org/10.1016/j.compgeo.2020.103806

Robert DJ (2017) A modified Mohr-coulomb model to simulate the behavior of pipelines in unsaturated soils. Comput Geotech 91:146–160. https://doi.org/10.1016/j.compgeo.2017.07.004

Schrefler BA, Scotta R (2001) A fully coupled dynamic model for two-phase fluid flow in deformable porous media. Comput Methods Appl Mech Eng 190:3223–3246. https://doi.org/10.1016/S0045-7825(00)00390-X

Tang CA, Liu H, Lee PKK et al (2000) Numerical studies of the influence of microstructure on rock failure in uniaxial compression—part I: effect of heterogeneity. Int J Rock Mech Min Sci 37:555–569. https://doi.org/10.1016/S1365-1609(99)00121-5

van Genuchten MT (1980) A closed-form equation for predicting the hydraulic conductivity of unsaturated soils. Soil Sci Soc Am J 44:892–898. https://doi.org/10.2136/sssaj1980.03615995004400050002x

Vanapalli SK, Fredlund DG, Pufahl DE, Clifton AW (1996) Model for the prediction of shear strength with respect to soil suction. Can Geotech J 33:379–392. https://doi.org/10.1139/t96-060

Vasarhelyi B (2003) Some observations regarding the strength and deformability of sandstones in dry and saturated conditions. Bull Eng Geol Environ 62:245–249. https://doi.org/10.1007/s10064-002-0186-x

Wang J, Zhang D, Wang N, Gu T (2019) Mechanisms of wetting-induced loess slope failures. Landslides 16:937–953. https://doi.org/10.1007/s10346-019-01144-4

Xie M, Esaki T, Cai M (2004) A time-space based approach for mapping rainfall-induced shallow landslide hazard. Env Geol 46:840–850. https://doi.org/10.1007/s00254-004-1069-1

Yu C, Tang S, Tang C et al (2019) The effect of water on the creep behavior of red sandstone. Eng Geol 253:64–74. https://doi.org/10.1016/j.enggeo.2019.03.016

Zhang LL, Fredlund DG, Zhang LM, Tang WH (2004) Numerical study of soil conditions under which matric suction can be maintained. Can Geotech J 41:569–582. https://doi.org/10.1139/t04-006

Zhou YF, Tham LG, Yan WM et al (2014) Laboratory study on soil behavior in loess slope subjected to infiltration. Eng Geol 183:31–38. https://doi.org/10.1016/j.enggeo.2014.09.010

Zienkiewicz OC, Humpheson C, Lewis RW (1975) Associated and non-associated visco-plasticity and plasticity in soil mechanics. Géotechnique 25:671–689. https://doi.org/10.1680/geot.1975.25.4.671

Zienkiewicz OC, Chan AHC, Pastor M et al (1997) Static and dynamic behaviour of soils : a rational approach to quantitative solutions. I. Fully saturated problems. Proc R Soc Lond A Math Phys Sci 429:285–309. https://doi.org/10.1098/rspa.1990.0061

Chapter 4
Numerical Simulations in FLAC^{3D}

Abstract This chapter begins by introducing the fundamental principles of the finite difference method (FDM), followed by a comparison with the finite element method (FEM) in terms of their respective advantages and limitations. It then presents a series of numerical simulations using the commercial software FLAC^{3D} to analyze rainfall infiltration and slope stability in a standard two-dimensional slope model. The discussion also covers the relationship between the permeability of weathered layers and rainfall intensity, and how this interaction affects slope stability under rainfall conditions. The focus then shifts to a comparative analysis of slope stability under rainfall using both two-dimensional (2D) and three-dimensional (3D) approaches. Special attention is given to the influence of slope geometry and groundwater levels on the variability of safety factors across different 2D cross-sections. These variations highlight the limitations of simplified two-dimensional analyses in capturing the complex geometric characteristics of slopes. The findings emphasize the importance of incorporating three-dimensional effects in slope stability assessments, particularly for heterogeneous slopes subjected to complex hydraulic conditions.

Keywords Numerical simulations · Uniaxial compression test · Slope surface undulation · Geometric features · 2D–3D comparison

4.1 Introduction

Numerical modeling techniques have become essential for addressing complex slope stability problems that are challenging or impossible to solve using conventional methods. These models are used to simulate both rock and soil slopes under various conditions, including complex geological and hydrological scenarios. Rock slopes, in particular, often involve numerous discontinuities such as joints, faults, and bedding planes. The accuracy of representing these discontinuities depends

Y. Jiang, X. Li, *Stability Evaluation of Slope Under Heavy Rainfall Based on Geometric Features*, Springer Natural Hazards,
https://doi.org/10.1007/978-981-95-8656-1_4

largely on the type of numerical model employed. Numerical methods for analyzing rock slope stability are typically categorized into three approaches including Continuum modeling, Discontinuum modeling and Hybrid modeling. Continuum modeling is best suited for slopes composed of massive, intact rock, weak rocks, or heavily fractured rock masses that behave like soil. In this approach, materials are treated as continuous throughout the model, with discontinuities represented as interfaces within the continuum. This method is effective for incorporating discrete fractures, such as faults and bedding planes, but it is less suitable for highly fractured rock masses. One commonly used numerical technique in continuum modeling is the finite difference method (FDM). FDM solves differential equations by approximating them with difference equations, where finite differences represent the derivatives. The problem domain is divided into smaller subdomains or elements, and the governing equations are solved numerically.

FDM-based continuum codes, such as $FLAC^{2D}$, $FLAC^{3D}$, Phase2, and PLAXIS, are widely used in slope stability analysis. These tools are capable of simulating complex slope behaviors, including groundwater effects, pore pressures, and dynamic interactions, provided the input parameters—such as constitutive models (elastic, elasto-plastic, creep), shear strength, groundwater characteristics, and in situ stress—are accurately defined. For two-dimensional continuum modeling, plane strain conditions are often assumed, which may not accurately represent slopes with heterogeneous structures, varying lithologies, or complex topographies. However, three-dimensional modeling tools like $FLAC^{3D}$ overcome these limitations by incorporating explicit finite-difference methods tailored for three-dimensional structures. $FLAC^{3D}$ represents materials using a polyhedral grid, which can deform and move in response to applied forces or boundary conditions. This tool uses an explicit Lagrangian calculation scheme and a mixed discretization zoning technique, allowing it to simulate plastic collapse and flow with high accuracy. An advantage of $FLAC^{3D}$ is its efficiency in handling large-scale three-dimensional calculations without excessive memory requirements, as no global stiffness matrices are formed. Additionally, the method's small time-step limitation and the need for damping are mitigated by automatic inertia scaling and damping mechanisms, which do not interfere with the failure modes being simulated. These capabilities make $FLAC^{3D}$ particularly effective for analyzing slopes under complex conditions, including large-strain deformations and material plasticity. Despite these advantages, certain limitations exist in $FLAC^{3D}$. Continuum modeling may not effectively capture the behavior of highly fractured or blocky rock masses, which are better suited to discontinuum or hybrid modeling approaches. Moreover, the accuracy and reliability of simulations depend on careful consideration of boundary effects, mesh quality, and computational resources, highlighting the importance of robust modeling practices.

In the last century, the evaluation of soil and rock mass behavior primarily relied on model experiments or limit equilibrium analysis. However, with advancements in computer technology and the emergence of the finite element method, numerical analysis has become more widely adopted in the field of geotechnical engineering. The finite element method approximates solutions by dividing the model into triangular or rectangular elements and solving for the response as a collection of these elements. Generally, a system of equations (global stiffness matrix) for elements connected at nodes is formed to satisfy deformation compatibility conditions, force equilibrium conditions, and stress-strain relationships. Additionally, the equilibrium conditions for the entire set of elements are used to determine the stress-deformation response. While the finite element method is the most commonly used technique for soil and rock mass analysis, many of the current finite element methods actively proposed are based on small deformation theory, which may not be suitable for representing large deformations (Chen et al. 2021).

Another widely used numerical simulation method is the finite difference method. This method approximates the differential terms of the fundamental differential equations using finite differences. By dividing the model into elements and approximating the acceleration terms in the equations of motion at each node, the stress-deformation behavior can be determined. Due to its simple formulation, explicit methods are frequently used in the calculation. When incrementally calculating stress and deformation over time, the finite difference method is often more advantageous than the finite element method (Zhang et al. 2015). When evaluating the behavior of the large-scale slope under rainfall conditions through numerical analysis, it is crucial to use methods based on large deformation theory. The application of large deformation theory has developed primarily within the field of structural analysis, addressing issues such as plastic deformation and ductile failure mechanisms. However, most of the previous research in soil and rock engineering regarding large deformation analysis remains within the framework of small deformation theory. In other words, existing analysis methods based on small deformation theory, such as the finite element method, struggle to clearly represent material nonlinearity, such as dilatancy and strain softening, as well as the geometric nonlinearity resulting from these factors. In contrast, finite difference analysis can account for both material and geometric nonlinearities and can perform computations based on large deformation theory. Therefore, this study uses the finite difference method, which is capable of considering both material and geometric nonlinearities.

Rainfall-induced slope instability is a critical topic in geotechnical engineering, necessitating accurate numerical simulation methods to predict and mitigate potential slope failures. Among the various numerical approaches available, the Finite Element Method (FEM) and the Finite Difference Method (FDM) are two widely utilized techniques. Each method has its advantages and limitations in the context of slope stability analysis under rainfall conditions. This discussion aims to provide an in-depth comparison of FEM and FDM, focusing on their application to simulate rainfall infiltration, pore pressure evolution, and slope deformation. FEM is known for its flexibility in handling complex geometries and material heterogeneity, making it particularly suitable for real-world slope problems. Its ability to accurately model irregular boundaries and varying material properties is a significant strength in slope analysis, where heterogeneity in soil layers and geometry often exists. FEM employs a mesh-based discretization approach, dividing the domain into smaller finite elements connected at nodes. This facilitates the incorporation of advanced constitutive models to simulate unsaturated soil behavior under rainfall. FEM can also seamlessly integrate unsaturated soil mechanics principles, such as those governing the soil-water retention curve (SWRC) and hydraulic conductivity variations with saturation. This capability is crucial in capturing the nonlinear effects of rainfall infiltration on slope stability.

Moreover, FEM offers a robust framework for coupled analyses, including fluid-solid interaction. The coupled hydro-mechanical models allow simultaneous simulation of changes in pore water pressure and stress-strain behavior, providing a more comprehensive understanding of rainfall-induced slope failures. However, FEM is computationally intensive, particularly for large-scale or three-dimensional problems. The generation of high-quality meshes is essential for accuracy, but this can be time-consuming and prone to errors in complex terrains. Additionally, the computational cost of FEM increases significantly with the complexity of the constitutive models or boundary conditions, posing challenges for large-scale simulations under limited computational resources. FDM differs from FEM in the following aspects: The mixed discretization is used in FDM to accurately model plastic collapse forces and plastic flow. Even for modeling static systems, FDM inherently uses a dynamic equation. The explicit method (a type of forward solver) is employed in FDM, allowing nonlinear problems to be solved in the same computation time as linear problems. In contrast, implicit methods in FEM would take significantly longer to solve nonlinear problems. Furthermore, an advantage of the explicit method in FDM is that there is no need to form a global stiffness matrix. This brings several benefits: A very large number of elements can be modeled using minimal memory. Since the global stiffness matrix is not used, large deformation problems can be numerically analyzed in nearly the same computation time as small deformation problems.

FDM is renowned for its simplicity and efficiency in solving partial differential equations. Unlike FEM, which requires a complex mesh, FDM employs a structured grid that simplifies the formulation and implementation. This advantage makes FDM particularly appealing for initial studies or problems with relatively simple geometries. FDM is well-suited for time-dependent problems, such as rainfall infiltration and its impact on slope stability. Its explicit or implicit time-stepping schemes enable efficient simulation of transient processes, such as the evolution of pore water pressure during prolonged rainfall. Additionally, FDM is computationally less demanding compared to FEM, making it suitable for large-scale or long-term simulations. The integration of rainfall infiltration models into FDM is straightforward, and many software tools based on FDM, such as FLAC^{3D}, are widely used in geotechnical engineering for slope stability analysis. These tools often include pre-built modules for rainfall infiltration, simplifying the implementation process for practitioners. Despite its strengths, FDM has notable limitations. The structured grid approach restricts its ability to handle complex geometries and material heterogeneity effectively. This limitation can lead to inaccuracies in simulating slopes with irregular boundaries or layered soil profiles. FDM also lacks the versatility of FEM in incorporating advanced constitutive models, particularly for unsaturated soils. Simplified assumptions may be required, reducing the model's accuracy in capturing nonlinear soil-water interactions. Additionally, FDM can encounter numerical instability issues, especially in problems involving steep slopes or high gradients in hydraulic conductivity.

FEM excels in accuracy and flexibility, particularly for problems requiring detailed representation of geometry and soil behavior. It is the preferred choice for advanced research applications, such as simulating complex slope failures involving multiple failure surfaces or intricate hydro-mechanical interactions. FDM, on the other hand, offers simplicity and computational efficiency, making it a practical choice for preliminary analyses or scenarios with simpler geometries. Its ability to handle transient processes efficiently makes it well-suited for analyzing the temporal evolution of slope stability under rainfall. Computational cost is a critical factor in selecting the appropriate method. FEM's higher accuracy comes at the expense of greater computational resources, while FDM provides a faster alternative with reduced precision in complex scenarios. Both FEM and FDM are valuable tools for simulating rainfall-induced slope instability, each with distinct strengths and weaknesses. FEM offers unparalleled flexibility and accuracy, making it ideal for complex and detailed analyses. FDM, with its simplicity and efficiency, serves as a practical alternative for time-sensitive or resource-constrained applications. A hybrid approach, leveraging the strengths of both methods, may offer the most robust solution for comprehensive slope stability analysis. Future advancements in computational power and numerical techniques may further enhance the applicability of these methods to address the challenges posed by rainfall-induced slope failures.

The large deformation finite difference analysis (FLAC^{3D} code: Fast Lagrangian Analysis of Continua in 3 Dimensions) is an explicit time-stepping analysis method developed by (Cundall and Board 1988). Explicit methods refer to a formulation where the solution at the "next time step" is directly obtained from the "current values" without solving an equation. Since obtaining a static solution for the problem at hand is essential, FLAC^{3D} employs viscous damping to achieve a pseudo-static analysis by obtaining the steady-state solution of dynamic response problems (static equilibrium or steady motion). This approach ensures computational stability when the modeled system is unstable. For nonlinear materials, physical instability is often inherently present (e.g., sudden collapse of a pillar). In reality, part of the system's strain energy serves as an energy source and is converted into dissipated kinetic energy. FLAC^{3D} directly models this process by including inertial terms. In contrast, analytical methods that do not include inertial terms require a specific mathematical procedure to handle physical instability. Even if this procedure successfully removes mathematical instability, the resulting path may not be realistic. In general, methods such as finite element analysis create a system of equations (global stiffness matrix) and solve them to obtain results, which is known as the "implicit method." On the other hand, the "explicit method" expresses the result for the time step $t + \Delta t$ directly from the result at time t based on the time increment Δt. According to this, in force-node displacement calculations using explicit methods, the calculations can be performed independently for each element, unaffected by neighboring elements.

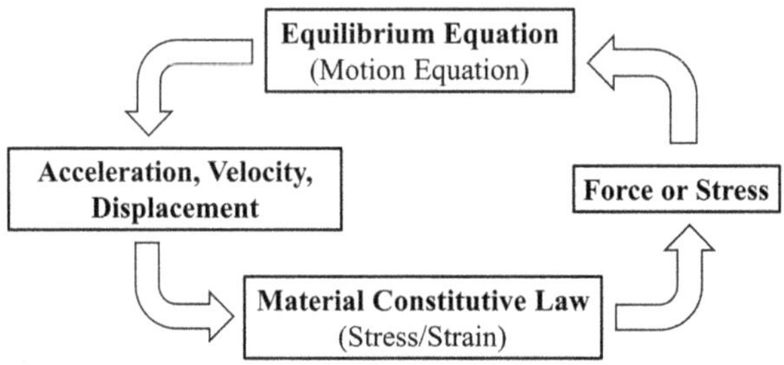

Fig. 4.1 Calculation cycle of the finite difference method

Figure 4.1 shows a typical computational cycle for the finite difference method. The procedure starts by solving the equations of motion using stresses (or forces) to obtain new velocities and displacements. Next, strain rates are derived from the velocities, and new stresses are obtained from the strain rates. Importantly, as long as control is within the bounds, the known values remain fixed, and the calculations based on these known values update the variables for all elements. For example, from previously computed velocities and displacements, new stresses or forces for each element are calculated. It is assumed that the velocity does not change within the boundary. In other words, the newly computed stresses do not influence the velocity.

The method may seem unreasonable from a general perspective, where the stress change at one point affects adjacent locations, altering the rate of change. However, by setting very small-time steps, it is assumed that physical information cannot be transmitted from one element to another within this time interval, and thus, physical values do not change in the adjacent elements during the time step. This implies that there is no need to create or store the global stiffness matrix frequently. For example, it eliminates the need to rebuild the global stiffness matrix when incorporating highly nonlinear material constitutive laws, such as strain softening. As a result, the explicit method can directly follow any nonlinear constitutive law, as shown in Fig. 4.2.

One important point to note is that since the explicit method assumes no change in unknown quantities between time step t and $t + \Delta t$, the time increment Δt must be as small as possible to prevent computational errors. The integration time step in explicit methods is determined by the Courant condition. In FLAC3D, the value of Δt follows a relationship derived from the minimum mass m and maximum stiffness k, and the optimal value is calculated based on this formula:

$$\Delta t \approx \sqrt{\left(\frac{m}{k}\right)} \tag{4.1}$$

In explicit methods, the governing equations need to be calculated repeatedly for small time increments, Δt. To enhance computational efficiency, discretization is applied not only in time but also in space, utilizing a finite difference method.

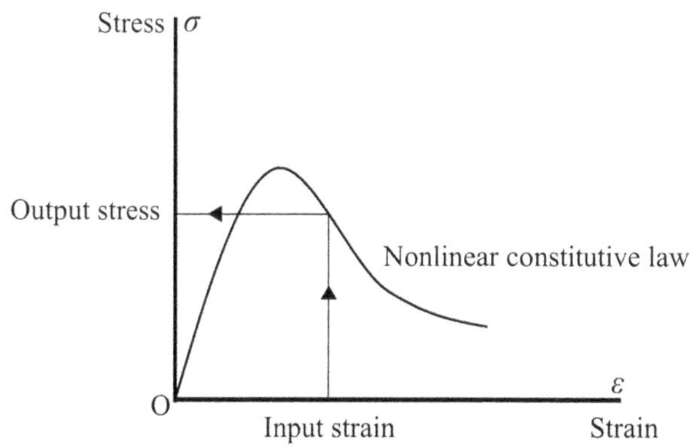

Fig. 4.2 Nonlinear constitutive laws by explicit method

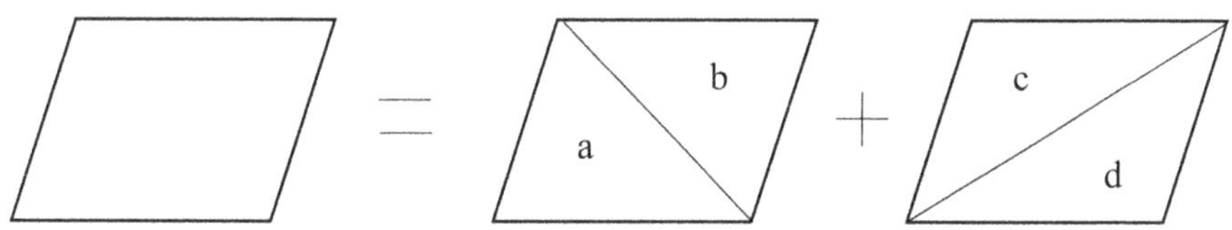

Fig. 4.3 Overlapping of triangular elements

Additionally, in the finite difference method, to effectively model the ground behavior leading to failure, particularly plastic flow, stress and strain are formulated using first-order interpolation, assuming they are constant within each element. However, this approach can lead to issues when modeling the deformation modes of rectangular elements. To address this, the finite difference method separates stress and strain into isotropic and deviatoric components, which are formulated using different methods. When considering two-dimensional rectangular elements, each element is defined as a pair of triangular elements, as shown in Fig. 4.3. The mixed discretization method assumes that the isotropic components remain constant across the entire rectangular element, while the deviatoric components are defined individually for each of the triangular elements (a, b, c, d).

The finite difference method, a core numerical solution in FLAC^{3D}, operates by approximating first-order spatial and temporal derivatives of a variable using finite differences. This approach assumes linear variations of the variable across finite spatial and temporal intervals. By employing these approximations, the governing equations of motion for a continuum are transformed into discrete forms of Newton's law at the nodes of the computational grid. The resulting system of ordinary differential equations is solved using an explicit finite difference scheme in the time domain. Spatial derivatives, essential for defining strain rates in terms of velocities, are derived based on the discretized medium. To accurately capture velocity variations and their corresponding spatial intervals, the continuum is divided into constant strain-rate elements with tetrahedral shapes. The vertices of these tetrahedra

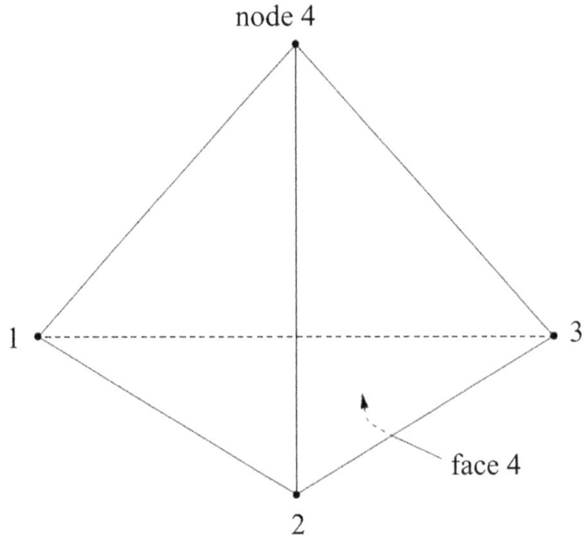

Fig. 4.4 Tetrahedron in FLAC^{3D}

correspond to the nodes of the computational mesh. A representation of such a tetrahedral element is illustrated in Fig. 4.4. This discretization and numerical approach enable FLAC^{3D} to model the dynamic behavior of the medium with high computational efficiency, making it well-suited for complex geotechnical and engineering analyses.

The finite difference formulation of the strain-rate tensor components for a tetrahedron is presented below as a foundation for deriving the nodal equations of motion. In each tetrahedral element, the nodes are locally numbered from 1 to 4, with the convention that face *n* is the face opposite to node *n* (refer to Fig. 4.4 for a visual representation). By applying the Gauss divergence theorem to the tetrahedron, the following expression can be derived:

$$\int_V v_{i,j} dV = \int_S v_i n_j dS \tag{4.2}$$

The integrals are taken over the volume and surface of the tetrahedron, respectively, where $[n]$ represents the outward unit normal vector to the surface. For a tetrahedron with constant strain rate, the velocity field varies linearly, and $[n]$ remains constant across each face. Consequently, upon performing the integration, Eq. (4.2) can be simplified to:

$$Vv_{i,j} = \sum_{f=1}^{4} \overline{v}_i^{(f)} n_j^{(f)} S^{(f)} \tag{4.3}$$

where the superscript (*f*) denotes the value of the associated variable on face *f*, and v_i represents the average value of the velocity component *i*. For a linear velocity variation across the tetrahedron, the relationship can be expressed as:

$$\overline{v}_i^{(f)} = \frac{1}{3} \sum_{l=1, l \neq f}^{4} v_i^l \tag{4.4}$$

where the superscript *l* corresponds to the value of the variable at node *l*. By substituting Eq. (4.4) into Eq. (4.3) and reorganizing the terms according to the contributions from each node, the following expression is derived:

$$V v_{i,j} = \frac{1}{3} \sum_{l=1}^{4} v_i^l \sum_{l=1, l \neq f}^{4} n_j^{(f)} S^{(f)} \tag{4.5}$$

If the v_i is replaced by 1 in Eq. (4.2), the following expression by applying the divergence theorem can be obtained:

$$\sum_{f=1}^{4} n_j^{(f)} S^{(f)} = 0 \tag{4.6}$$

Using this last relation and dividing Eq. (4.3) by *V*, the following expression can be obtained:

$$u_{i,j} = -\frac{1}{3V} \sum_{l=1}^{4} v_i^l n_j^{(l)} S^{(l)} \tag{4.7}$$

and the components of the strain-rate tensor can be expressed as:

$$\xi_{ij} = -\frac{1}{6V} \sum_{l=1}^{4} \left(v_j^l n_j^{(l)} + v_j^l n_i^{(l)} \right) S^{(l)} \tag{4.8}$$

4.2 Analysis of Standardized Slopes

Field investigations have shown that most slope failures in Japan occur within permeable residual soil layers overlying relatively competent rock foundations (Cho et al. 2021). Their research reported that failure depths typically do not exceed 2 m and that slope angles commonly range between 30° and 50°. Because the present study focuses on shallow failure mechanisms, the slope geometry was simplified by adopting a representative inclination of 45° and assigning a 5 m thick weathered soil layer above the bedrock (Fig. 4.5). The model height and length were set to 24 m and 34 m, respectively. A mesh size of 0.5 m was used, resulting in 2445 zones and

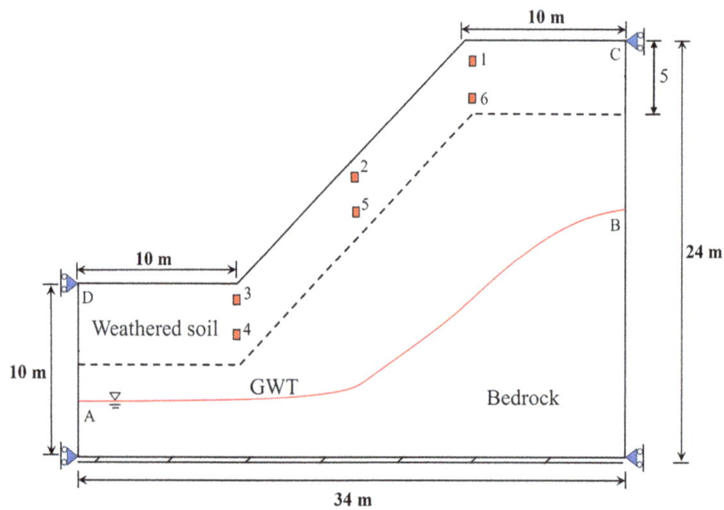

Fig. 4.5 The calculation model of slope (Li et al. 2025)

Table 4.1 The mechanical parameters of slope in a saturated state

	Friction angle °	Cohesion kPa	Poisson's ratio	Young's modulus MPa	Permeability m/s	Density (Dry) kg/m³
Colluvial soil	30	20	0.35	35	6×10^{-6}	1800
Bedrock	36	100	0.13	300	5×10^{-7}	2200

4831 grid points. The geotechnical properties of both the colluvial soil and the bedrock are summarized in Table 4.1. Rainfall infiltration was applied along the slope surface. To examine the stability differences between homogeneous and heterogeneous slopes under short-duration heavy rainfall, the lateral boundaries above the groundwater table (AD and BC) were treated as impermeable. Variations in groundwater level are not regarded as a key factor influencing the failure mechanism. Accordingly, the groundwater table was assumed to remain constant throughout the numerical simulations, as shown in Fig. 4.5.

The numerical simulation procedure is summarized in Fig. 4.6. A negative pore pressure of −150 kPa was applied along the slope surface to establish an initial seepage equilibrium. This produced a linearly increasing saturation profile from the groundwater level to the ground surface (Fig. 4.7a). In the simulation, a sustained rainfall intensity of 20 mm/h was adopted. This value corresponds to the average maximum hourly precipitation recorded in Sasebo City during the period of 13–17 September 2023. When rainfall intensity slightly exceeds the saturated hydraulic conductivity of the soils, additional rainfall generates surface runoff rather than further infiltration. As shown in Fig. 4.7b, the modified unsaturated permeability coefficient for the shallow layer ranges from approximately 1.0×10^{-6} to

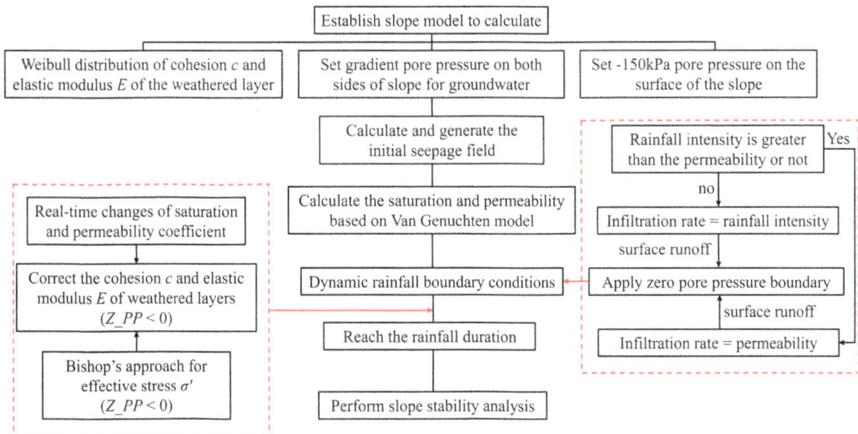

Fig. 4.6 The flowchart of the numerical simulation

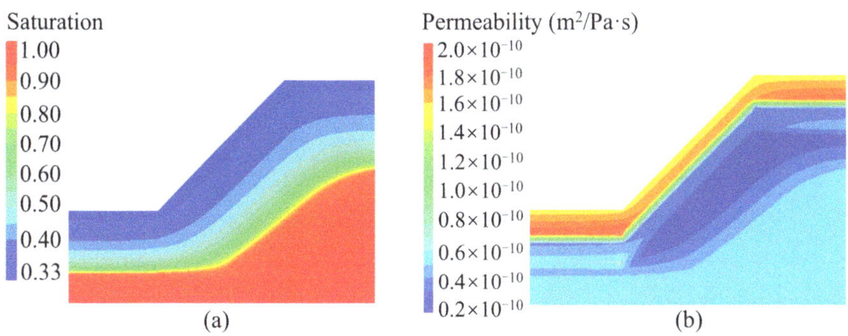

Fig. 4.7 Distribution of saturation (**a**) and permeability (**b**) in the slope before rainfall (Li et al. 2025)

2.0×10^{-6} m/s. This range is consistent with laboratory measurements obtained from soil samples collected at the Sasebo landslide site, supporting the reliability of Eq. (3.4). Figure 4.8 illustrates the evolution of the saturation field with increasing rainfall duration. The unsaturated zone progressively diminishes as rainfall continues, which corresponds to a decline in the safety factor (Fig. 4.9). Because weathered soils exhibit higher permeability than the underlying bedrock, their shallow layers experience a rapid reduction in unsaturated area during short-term rainfall events. To more accurately represent rainfall infiltration, both real-time variations in permeability and dynamic rainfall boundary conditions were incorporated into the simulation. Prior to the onset of surface runoff, the rainfall boundary was modeled as a flow boundary. Once runoff developed, the boundary condition was switched to a zero–pore pressure condition. A detailed analysis of the Sasebo landslide will be provided in the next chapter.

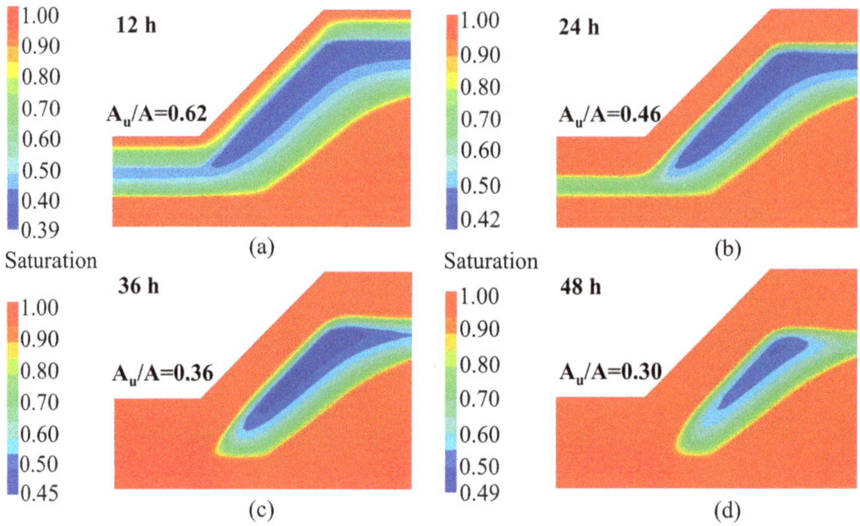

Fig. 4.8 Distributions of saturation of the slope after 12 h (**a**), 24 h (**b**), 36 h (**c**) and 48 h (**d**) of rainfall (A_u is the area of the unsaturated zone, A is the total area) (Li et al. 2025)

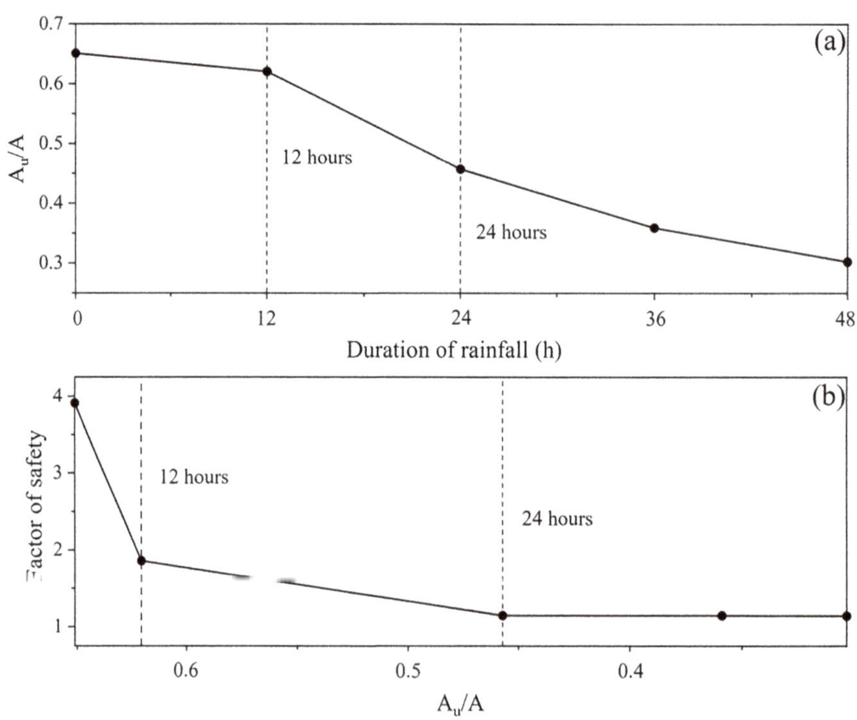

Fig. 4.9 (**a**) Variation of unsaturated areas during rainfall infiltration. (**b**) Relationship between factor of safety and unsaturated areas during rainfall infiltration

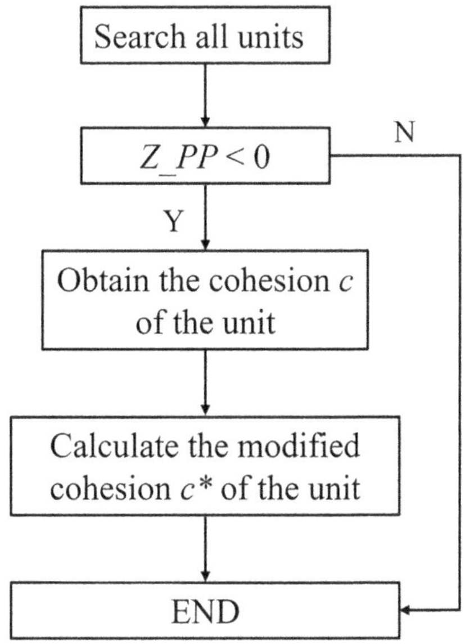

Fig. 4.10 The program chart of correcting cohesion in the unsaturated zone

To clearly demonstrate the influence of matric suction on cohesion ccc and elastic modulus E, a homogeneous slope was used in the numerical simulations. Because the analysis focuses on shallow slope failure, suction-related corrections for c and E were applied only to the weathered soil layer. The procedure for correcting cohesion in the unsaturated zone during rainfall follows Eq. (3.30), and the correction of elastic modulus is based on Eq. (3.31). The detailed implementation procedure in FLAC^{3D} is presented in Fig. 4.10. The initial cohesion and elastic modulus of the weathered soil were 20 kPa and 35 MPa, respectively. When matric suction was incorporated, cohesion increased to between 27 and 35 kPa, and the elastic modulus rose to between 75 and 110 MPa (Fig. 4.11). These results clearly show that matric suction substantially enhances both the shear strength and stiffness of the weathered soil.

The fundamental mechanism of slope failure under rainfall infiltration is governed by changes in pore-water pressure that modify the effective stress state. A reduction in effective stress may trigger plastic yielding. In this study, the shear strength of partially saturated soil is evaluated using the modified Mohr–Coulomb criterion, which incorporates effective stress under partially saturated conditions as expressed in Eq. (3.27). Effective stress is computed using Bishop's parameter χ, a saturation-dependent coefficient that accounts for the role of matric suction in enhancing shear strength. Experimental evidence from triaxial tests on unsaturated soils indicates that the internal friction angle φ remains essentially constant, whereas cohesion c decreases with matric suction during both shear and wetting processes (Wang et al. 2019). Consequently, a constant φ is adopted in this study. This treatment is consistent with Bishop's effective stress principle, which attributes the contribution of matric suction solely to apparent cohesion. The shear modulus G and

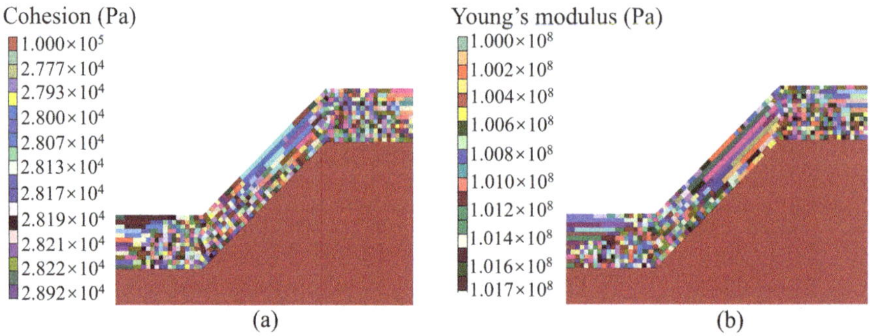

Fig. 4.11 Cohesion c (**a**) and the Young's modulus E (**b**) of the slope before rainfall (Pa) (Li et al. 2025)

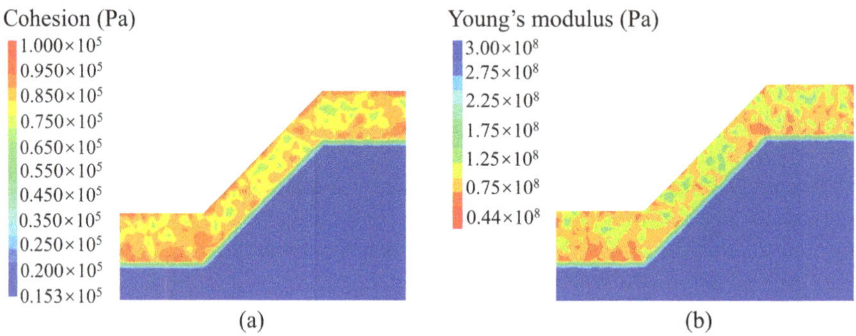

Fig. 4.12 Weibull distribution of (**a**) cohesion c and (**b**) elastic modulus E (Pa) (Li et al. 2025)

Young's modulus E are modified to reflect partial saturation of weathered soils. These relationships, calibrated against experimental data, are incorporated into the numerical model to capture stiffness degradation as the soil transitions from unsaturated to saturated conditions (Hoyos et al. 2015). As shown in Eq. (3.31), modulus values decrease with increasing saturation, reflecting the softer mechanical response of the soil under wet conditions. The reference values σ_0 for Young's modulus E and cohesion c used in Eq. (3.2) are 35 MPa and 20 kPa, respectively. These values represent the average saturated Young's modulus and cohesion obtained from multiple triaxial compression test datasets. Figure 4.12 presents the Weibull distributions of cohesion and Young's modulus for the weathered layers, incorporating matric suction effects ($m = 3$). These distributions were generated using the built-in random distribution function in FLAC^{3D} and derived inversely from Eq. (3.2). As a result, both c and E follow Weibull-type distributions, as shown in Fig. 4.13.

This chapter evaluates the influence of rainfall on slope stability primarily through variations in the factor of safety (FoS), rather than through small changes in displacement when the FoS remains above 1. The results show that the FoS computed with heterogeneous parameters is consistently lower than that obtained using

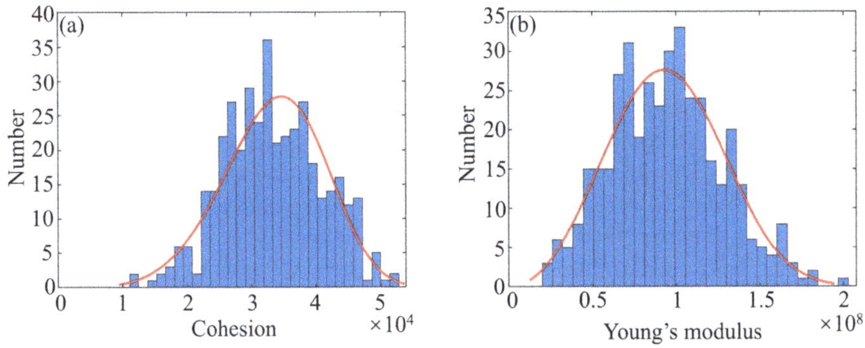

Fig. 4.13 The Weibull distribution of (**a**) cohesion *c* and (**b**) elastic modulus *E* (Pa) (Li et al. 2025)

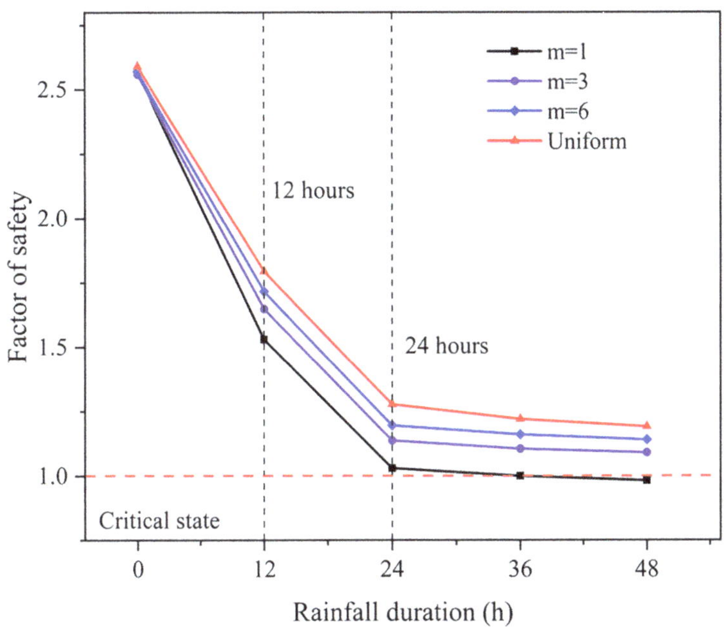

Fig. 4.14 Comparison of FoS between heterogeneous and uniform slopes during rainfall (Li et al. 2025)

homogeneous parameters under the same rainfall duration (Fig. 4.14). A higher m value corresponds to a higher FoS, indicating that increased material non-uniformity reduces slope stability during rainfall infiltration. The reduction in stability is most pronounced for the highly heterogeneous condition ($m = 1$), where the FoS drops below 1 after 48 h of rainfall (Fig. 4.14). This clearly suggests that greater heterogeneity (lower m) substantially weakens the slope. A lower m value reflects greater variability in material properties, such as cohesion and elastic modulus, which can generate localized weak zones. These weak zones may serve as potential failure surfaces, significantly reducing the safety factor as rainfall progresses. The notable degradation in stability for $m = 1$ underscores the importance of accurately representing material heterogeneity in slope stability analysis. Assuming a homogeneous weathered layer (higher m) may underestimate the likelihood of failure under prolonged rainfall. This finding highlights the critical need to incorporate realistic spatial variability of weathered-layer properties into numerical models to ensure reliable assessments of rainfall-induced slope stability.

To analyze rainfall infiltration within the slope, six monitoring points were installed in the numerical model to record pore-water pressure variations during rainfall (Fig. 4.5). Monitoring points 1–3 were placed 2 m below the slope surface, and points 4–6 were positioned at a depth of 4 m. In this simulation, the shape parameter m in the Weibull distribution was set to 1. The results show that the negative pore-water pressure at a depth of 2 m rose rapidly and became positive after 12 h of rainfall (Fig. 4.15a). After 24 h, the negative pore pressure at a depth of 4 m also transitioned to positive values (Fig. 4.15b), marking the wetting front reaching this deeper layer. This transition corresponds to the notable decline in the slope's factor of safety during the initial 24 h of rainfall (Fig. 4.14). The dissipation of matric suction, which initially contributes to shear strength, is a key factor driving this reduction in stability. As suction dissipates, the effective stress decreases, leading to a loss of shear strength. This process is particularly evident in the upper portions of the slope, where infiltration proceeds more rapidly. These findings highlight the critical influence of rainfall-induced pore-pressure evolution on slope stability and emphasize the necessity of considering the temporal behavior of pore-water pressure in numerical analyses. Overall, the results underscore how the loss of matric suction significantly accelerates the reduction in slope stability.

4.2.1 Permeability and Rainfall Intensity

After the shallow layer of the soil reaches saturation rapidly, most of the rainfall no longer infiltrates but forms surface runoff on the slope. The thin water film created on the slope generates an equivalent hydraulic head to the elevation. In such cases, the safety factor of the soil slope may exhibit minimal changes, or even remain constant. Many studies have extensively explored the impact of the relationship between saturated hydraulic conductivity and rainfall intensity on slope stability during rainfall infiltration (Ng and Shi 1998; Cho and Lee 2001; Cai and Ugai 2004; Wang et al. 2023). When the rainfall intensity is of the same order of magnitude as the saturated hydraulic conductivity of the rock-soil mass and slightly exceeds it, the slope safety factor experiences the most pronounced decrease (Ng and Shi 1998; Cho and Lee 2001). Considering the varying hydraulic conductivities resulting from different rock types constituting the slope, this study selected three typical hydraulic conductivities for numerical simulations to validate the influence of the relationship between rainfall intensity and saturated hydraulic conductivity on slope stability

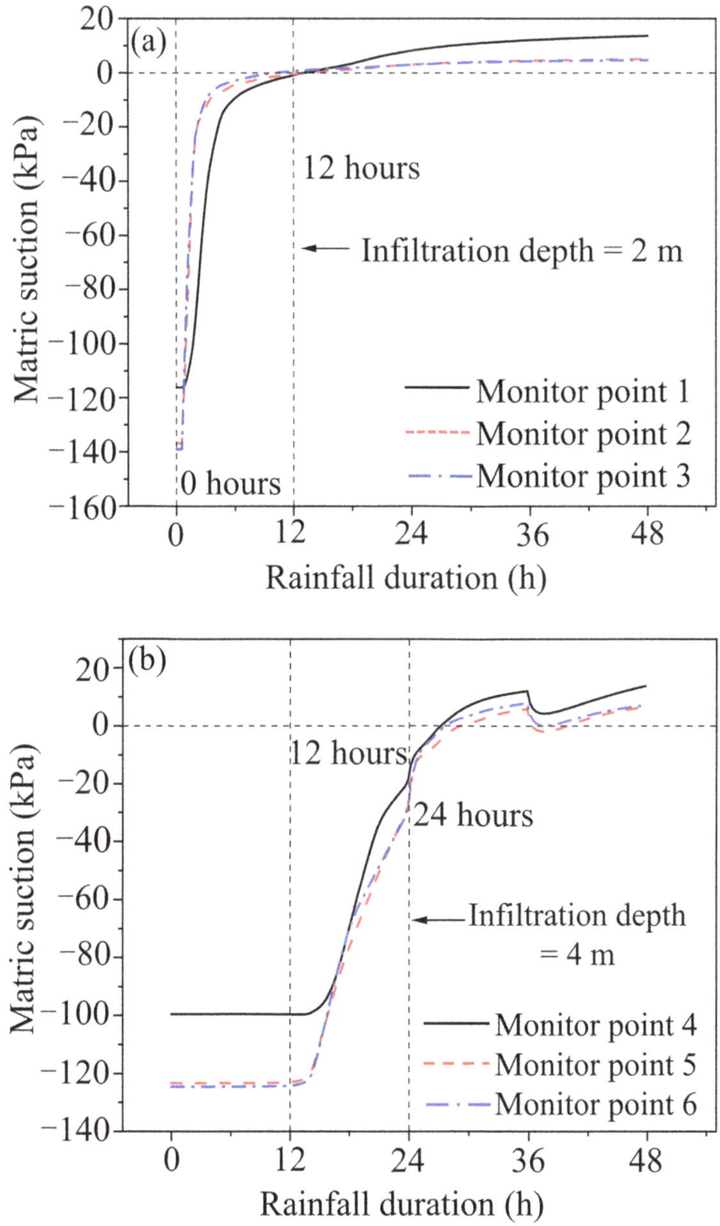

Fig. 4.15 Variation of matric suction during rainfall infiltration process: (**a**) Monitor points: 1, 2, 3. (**b**) Monitor points: 4, 5, 6 (Li et al. 2025)

Table 4.2 Permeability of different kinds of rocks

Rock type		Permeability (weathered layers) m/s
Igneous rocks	Basalt	5×10^{-7}
	Granite	$5.2 \times 10^{-5} \sim 3.3 \times 10^{-6}$
Sedimentary rocks	Mudstone	2×10^{-7}
	Sandstone	5×10^{-6}
	Siltstone	1.76×10^{-5}
Metamorphic rocks	Slate	$10^{-4} \sim 10^{-6}$
	Schist	

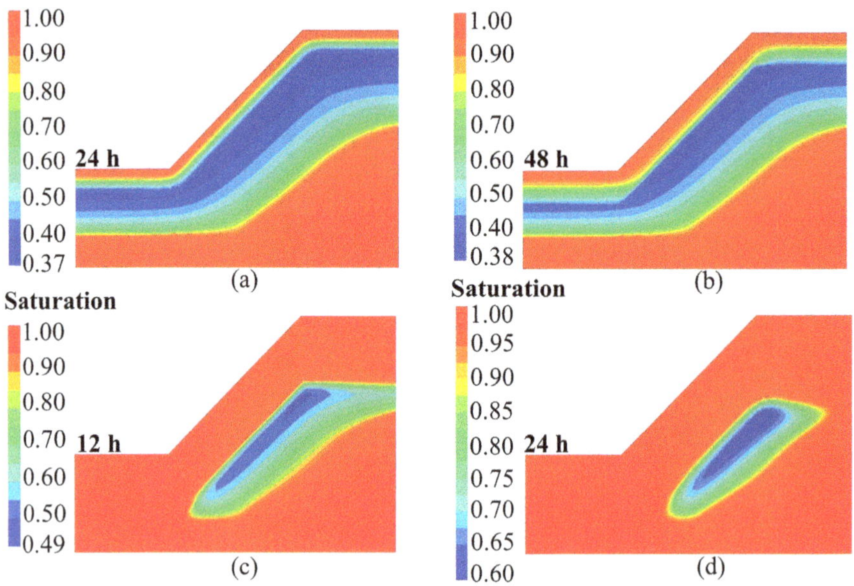

Fig. 4.16 The distribution of saturation in the slope after 24 h (**a**) and 48 h (**b**) of rainfall (Case 1). The distribution of saturation in the slope after 12 h (**c**) and 24 h (**d**) of rainfall (Case 3). The distribution of saturation in the slope of case 2 after rainfall can be found in Fig. 4.8

(Table 4.2). Since this study primarily focuses on the weathered layers of the slope, the permeability characteristics of the bedrock will be explored in future research. For the purposes of the current analysis, the permeability coefficient of the bedrock is uniformly set at 5×10^{-7} m/s across all three scenarios. As shown in Fig. 4.16, it is evident that the permeability coefficient of the weathered layer directly determines the rate of rainfall infiltration. When the permeability coefficient is relatively small (Case 1), the unsaturated zone decreases at a slower rate after 24 and 48 h of rainfall. In contrast, when the permeability coefficient is larger (Case 3), the unsaturated zone of the slope decreases rapidly after 12 and 24 h of rainfall (Table 4.3).

Table 4.3 Permeability of weathered layers in the slope

Calculation cases	Permeability of weathered layers(m/s)
1	$k_1 = 6.0 \times 10^{-7}$
2	$k_2 = 6.0 \times 10^{-6}$
3	$k_3 = 3.0 \times 10^{-5}$

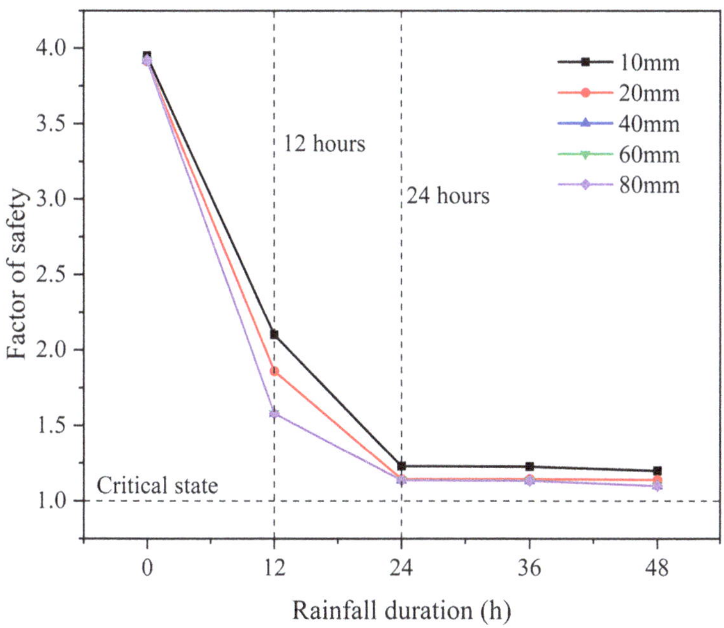

Fig. 4.17 Relationship between factor of safety and rainfall hours (Case 2)

In this study, three typical permeability coefficient values were selected based on the permeability of different weathered rocks for calculations. Rainfall intensities were set at 10 mm/h, 20 mm/h, 40 mm/h, 60 mm/h, and 80 mm/h. When the corrected unsaturated permeability coefficient is significantly smaller than the rainfall intensity, for instance, in Case 1 where the permeability coefficient of highly weathered layer is $k_1 = 6 \times 10^{-7}$ m/s, and the infiltration rate for 20 mm/h rainfall is 5.56×10^{-6} m/s, slope surface runoff occurs in the early stages of rainfall. Further increasing the rainfall intensity does not alter the seepage state within the slope, resulting in consistent safety factor changes across different rainfall intensities (Fig. 4.17). When the permeability coefficient increases to 6×10^{-6} m/s, which is between the infiltration rates of 20 mm/h and 40 mm/h (11.12×10^{-6} m/s), it is observed that the safety factor decreases slightly as the rainfall intensity increases from 10 mm/h to 40 mm/h. However, with a continued increase in rainfall intensity, the trend of safety factor changes becomes constant. As the permeability coefficient further increases to 3.0×10^{-5} m/s (Case 3), surpassing the infiltration rate of

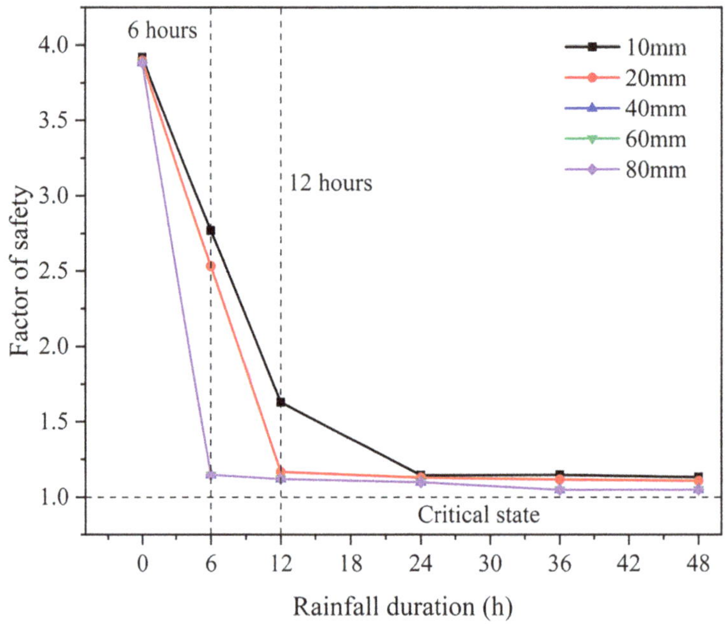

Fig. 4.18 Relationship between factor of safety and rainfall hours (Case 3)

80 mm/h (2.2×10^{-5} m/s), the safety factor decreases continuously with the increase in rainfall intensity from 10 mm/h to 80 mm/h. The time of abrupt decrease in safety factor also reduces from 12 h to 6 h. The decrease in the factor of safety (FOS) to around 1, as shown in Fig. 4.18. This is attributed to the initial cohesion c and internal friction angle φ, along with the geometry of the slope, determining the strength reserve of the slope. On the other hand, no leakage boundary conditions were set for the cases, and above the groundwater table, impermeable boundaries were applied on both sides of the slope in the current computations. Therefore, with a sufficiently high permeability coefficient, as the rainfall intensity increases, the safety factor will decrease more rapidly and to a lower level. If both sides were considered permeable boundaries, an increase in the permeability coefficient would imply a faster drainage of rainwater. Consequently, continuously increasing rainfall intensity might not lead to a significant additional decrease in the safety factor. However, the study of boundary conditions is not the focus of this research, and detailed exploration of this aspect will be undertaken in subsequent studies. Even after saturation due to rainfall, the safety factor of the slope remains around 1. This outcome is a result of the artificial nature of the numerical simulation model used in this study. The subsequent research will delve into the impact of slope soil and rock strength parameters, as well as geometry, on slope stability under rainfall conditions. According to the simulation results, it is evident that with a higher permeability coefficient of the slope, larger rainfall amounts lead to a faster decrease in the safety factor of slopes over time.

4.2.2 *Uniaxial Compression Test in FLAC^{3D}*

The uniaxial compressive strength (UCS) of rock is a crucial parameter that governs the mechanical behavior and failure characteristics of rocks under loading conditions (Yarali and Kahraman 2011). The UCS of rocks has been widely used for rock mass classification, stability analysis, and engineering design in rock engineering (Fu et al. 2022). Over the past few decades, extensive experiments and numerical simulations have been conducted using uniaxial compression tests to investigate mechanical failures of rocks (Xie et al. 2024). These studies aimed to understand how materials respond to compressive stress by examining factors such as strength, deformation characteristics, failure modes. This study employs numerical simulations of uniaxial compression tests to investigate the impact of different porosities on the peak strength of rock samples under varying saturation conditions, as well as the effect of heterogeneity on peak strength of rock samples. To simulate the deformation behavior of rock under loading, a standard rock specimen model with dimensions of 50 mm × 100 mm was created and meshed with a 1 mm grid size in FLAC^{3D}, as illustrated in Fig. 4.19a. Figure 4.19b illustrates the Weibull distribution of the cohesion c for the rock samples. The mechanical properties of the sandstone specimen are detailed in Table 4.4. Axial velocities were applied to the model to simulate uniaxial compression tests. These tests were performed on rock specimens with varying saturation levels, as shown in Table 4.5. The physical and mechanical parameters of the rock specimens, including internal friction angle φ, cohesion c, elastic modulus E, and Poisson's ratio μ, were reduced according to the relationship between porosity and weakening coefficient λ. The specific reduction coefficients corresponding to different porosities are listed in Table 4.5. The selection of the reduction factor λ is primarily based on the experimental data provided by Pan et al. (2020). It is worth noting that the exponential relationship between the reduction factor and the saturation level of the specimen was derived under controlled laboratory conditions. In actual rock slopes, the changes in saturation under rainfall

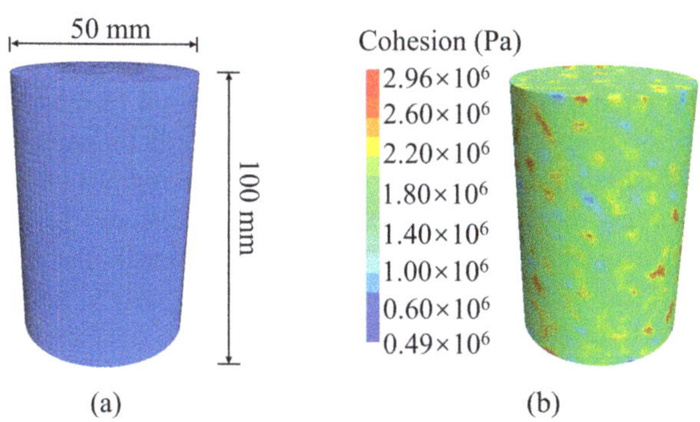

Fig. 4.19 (**a**) Model of the rock specimen. (**b**) Weibull distribution of cohesion c (Pa) of the rock specimen

Table 4.4 The material properties of rock samples under dry condition

Properties	Friction angle °	Cohesion MPa	Poisson's ratio	Density kg/m^3	Young's modulus GPa
Rock samples	40	7.2	0.22	2500	2.5

Table 4.5 Three groups of rock samples in the numerical simulations of USC

Group	Condition	Porosity (n)	Weakening coefficient (λ)
1	100% saturated	0.14	0.6
	50% saturated		0.8
	Dry		1
2	100% saturated	0.16	0.3
	50% saturated		0.65
	Dry		1
3	100% saturated	0.20	0.1
	50% saturated		0.55
	Dry		1

conditions are significantly less pronounced compared to those observed in laboratory settings.

To investigate the impact of heterogeneity in the physical and mechanical parameters of rock specimens on the peak strength in uniaxial compression tests, each of the originally set three test groups was further divided into four subgroups with varying degrees of homogeneity based on different shape parameters *m*. These subgroups were classified as $m = 1$, $m = 3$, $m = 6$, and a homogeneous group. The results reveal that for homogeneous rock samples with same porosities, the peak strength σ_{peak} decreases as the parameters including cohesion *c*, friction angle φ, Young's modulus *E*, and Poisson's ratio μ were reduced according to the degree of saturation (Fig. 4.20). As shown in Fig. 4.21, a smaller porosity results in a higher peak strength σ_{peak} for homogeneous rock samples with the same degree of saturation. When the porosity and saturation level are the same, the peak strength σ_{peak} of the rock samples decreases with a decreasing heterogeneity index *m*, and vice versa (Fig. 4.22). From the maximum shear strain contour plot in Fig. 4.23a, it can be observed that when heterogeneity degree of the rock samples is $m = 1$, a distinct continuous shear band was not shown during the USC tests. Instead, some localized areas of high shear strain are present. As the degree of heterogeneity increases to $m = 3$, which indicates a more uniform rock sample, the regions of maximum shear strain start to become more concentrated and show a tendency towards forming a continuous shear band (Fig. 4.23b). When the rock sample becomes even more uniform with $m = 6$, continuous shear bands are clearly visible in Fig. 4.23c, and a distinct continuous shear band is observed at the end of the test for the homogeneous rock sample, as shown in Fig. 4.23d. The physical scale of the rock samples in laboratory tests with standardized dimensions differs significantly from the scale of the actual slope represented in the numerical model. This disparity can affect the interpretation of material properties, such as strength and deformability, because

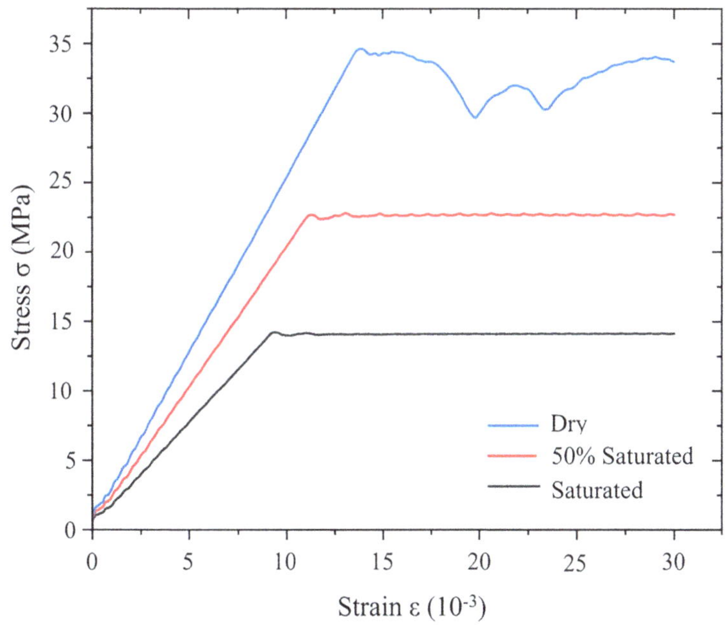

Fig. 4.20 Stress-strain plots comparison for samples with different degrees of saturation (n = 0.14)

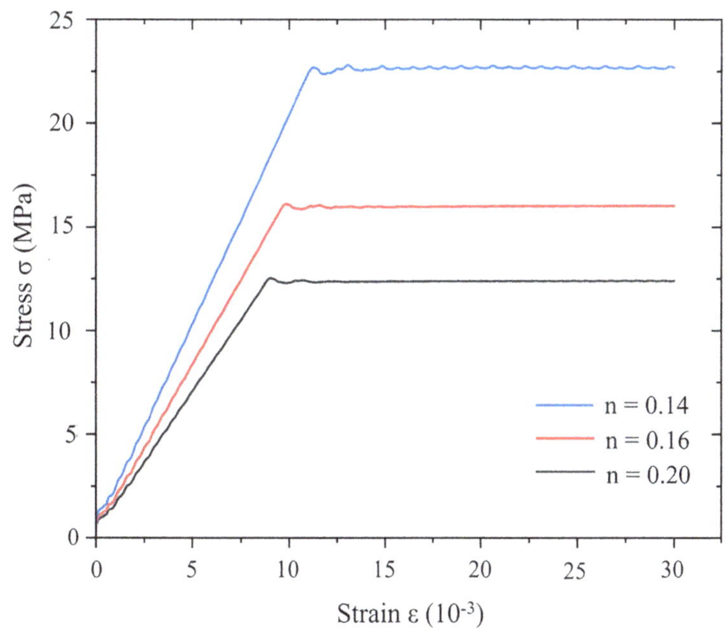

Fig. 4.21 Stress-strain plots comparison for samples with different porosity (50% Saturated)

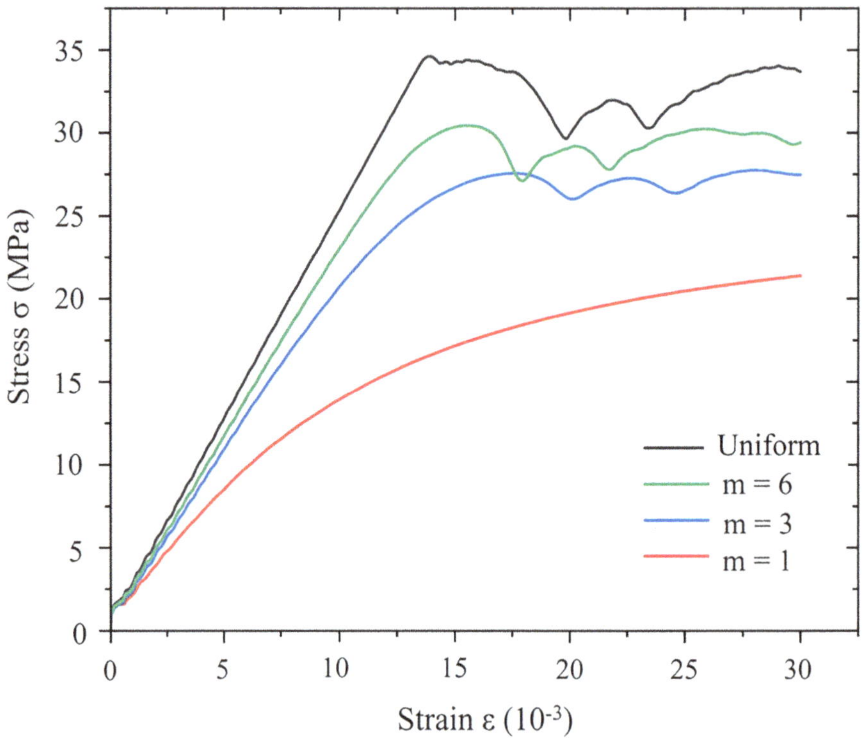

Fig. 4.22 Stress-strain plots comparison for samples with different degrees of heterogeneity (n = 0.14)

laboratory samples might not capture the inherent heterogeneity, and discontinuities present in the slope-scale material. However, this chapter mainly focuses on the effects of different degrees of heterogeneity and porosity on the peak strength of rock samples, as well as the influence of heterogeneity on the formation of shear bands during the Uniaxial Compression Test. Therefore, the obtained uniaxial compression strength was not used in the numerical calculations of the 2D slope. These two calculations are separate and are both aimed at exploring the impact of rock heterogeneity and water-induced weakening effects on rock strength and slope stability. Future research will be conducted on large-scale uniaxial compression laboratory tests to investigate the effects of heterogeneity and porosity on the strength of soft rocks.

4.2.3 Two-Dimensional Rock Slope Model

This study conducted a reproduction analysis on the slope failure that occurred in July of 2009 in Benjo, Fukuchi Town, Fukuoka Prefecture, Japan. The simplified slope model features highly and moderately weathered layers with the bedrock

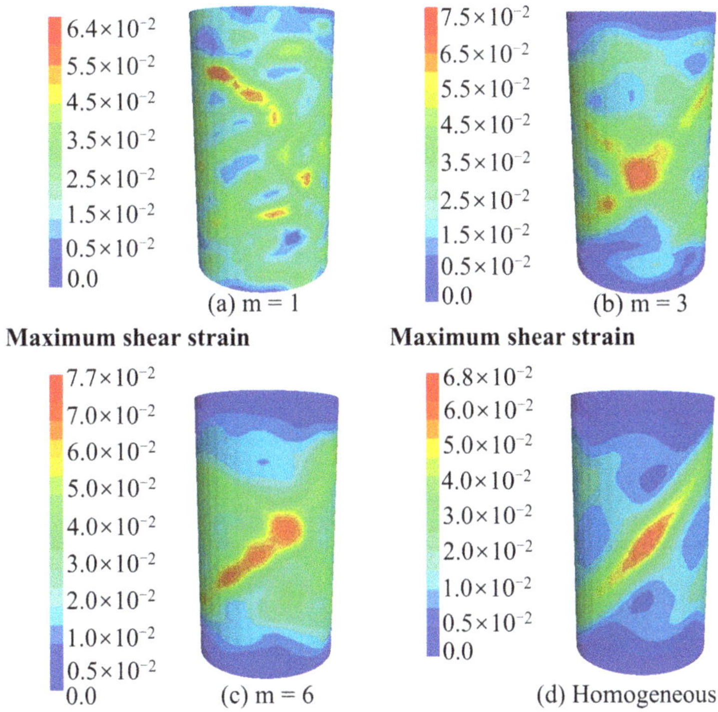

Fig. 4.23 Fracture pattern comparison for compression of Mohr-Coulomb samples with different degrees of heterogeneity (n = 0.14)

composed of sandstone, as shown in Fig. 4.24. The height and length of the slope model were set to 26 m and 36 m, respectively. The slope model was divided into 2568 zones and 5376 gridpoints. The parameters of the slope are based on the report by the Kyushu Northern Japan Landslide Disaster Investigation Team of the Japanese Geotechnical Society. Table 4.6 lists the physical mechanical parameters of the slope used in the numerical analysis. The collapse pattern was a surface failure occurring in the strongly weathered layer after two consecutive days of heavy rainfall. The boundaries on both sides of the slope above the groundwater level (AD and BC) were simply set as impermeable boundaries. As this study concentrates on the effects of short-term heavy rainfall on the shallow layers of slopes, the influence of groundwater level is not a primary concern in the analysis of failure mechanisms. The initial groundwater level (AB) was assumed to be relatively low in the slope, as shown in Fig. 4.24. The bottom of slope model was constrained in both vertical and horizontal displacements, and both sides were constrained in horizontal displacement. The initial saturation levels of the strongly weathered layer were set to 0.33 based on field investigations. According to the soil-water characteristic curve (SWCC) fitted using the Van Genuchten model (van Genuchten 1980), a seepage equilibrium calculation can generate a linearly increasing negative pore pressure field from the groundwater level to the slope surface by setting a negative pore

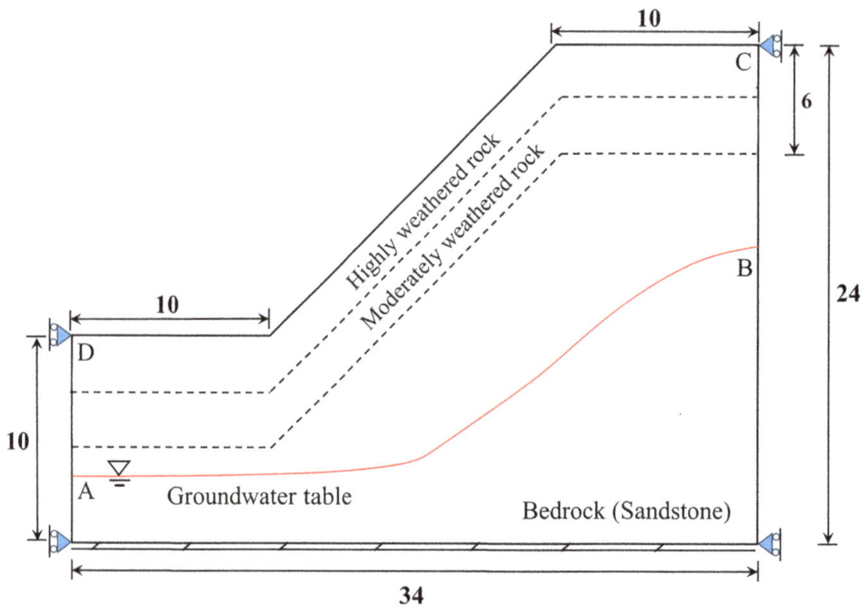

Fig. 4.24 The calculation model of slope (unit: m)

Table 4.6 The mechanical parameters of slope

	Friction angle °	Cohesion kPa	Poisson's ratio	Density kg/m³	Young's modulus MPa	Permeability m/s
Highly weathered	35	20	0.30	1800	200	6×10^{-6}
Moderately weathered	38	30	0.20	2000	300	1×10^{-6}
Bedrock	40	100	0.13	2100	600	5×10^{-7}

pressure of −100 kPa on the surface of the slope. The rainfall intensity was set to 20 mm/h and input as a velocity head at the nodes on the slope surface. For detailed information on the unsaturated seepage calculations of a two-dimensional slope in FLAC^{3D}, please refer to the paper published by us (Li et al. 2024). The reduction coefficient λ for the water-induced weakening effect on rocks depends on the porosity of rocks. In this study, the average porosity of the strongly weathered layer is approximately 0.16, while the porosity of the moderately weathered layer is about 0.14 based on field surveys. Since the weathered rock samples were taken during the rainfall period, the measured values of cohesion c and internal friction angle φ remained at a certain degree of saturation. The reduction coefficients λ after 48 h of continuous rainfall at a rate of 20 mm/h are 0.7 for highly weathered layers and 0.8 for moderately weathered layers based on the parameters in Table 4.5. As noted in the study by Iyare et al. (2023), there is a difference between the saturation levels achieved in situ and those achieved through prolonged immersion in laboratory

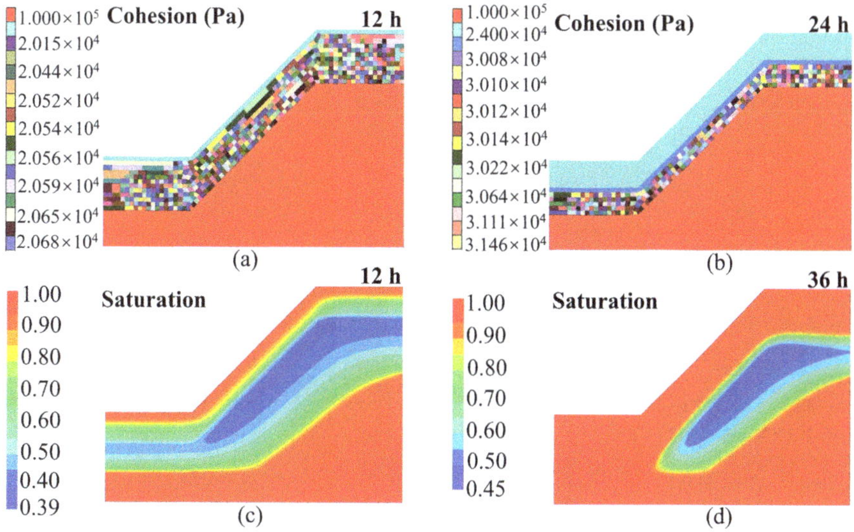

Fig. 4.25 The change of cohesion c after (**a**) 12 h to (**b**) 24 h of rainfall, the change of saturation after (**c**) 12 h to (**d**) 36 h of rainfall

conditions. Therefore, the reduction coefficients λ obtained is more conservative in this numerical simulation compared to the reduction coefficients for saturated sandstone samples obtained from uniaxial tests in the study by Pan et al. (2020). Figure 4.25 shows the variation of cohesion c for two weathered layers after 12 and 24 h of rainfall, considering the increasing of cohesion in the weathering layers due to matric suction. As shown in Fig. 4.25b, it can be observed that the cohesion of the first weathered layer reduces to 16 kPa after saturation. Compared to its initial state of 20 kPa, this represents a reduction of 80% which is attributed to the water-induced weakening of the rocks.

Different from the method of Chap. 3, which applied Bishop's effective stress principle to modify the cohesion c in the unsaturated regions of the slope, the principles of the shear strength of unsaturated soils proposed by Fredlund et al. (1978) was adopted to correct the cohesion of weathered layers in the unsaturated zone in this section, as shown in Eq. (4.9):

$$\tau = c' + \left(\sigma - u_{\mathrm{a}}\right)_{\mathrm{f}} \tan\varphi' + \left(u_{\mathrm{a}} - u_{\mathrm{w}}\right)_{\mathrm{f}} \tan\varphi^{\mathrm{b}} \tag{4.9}$$

where τ is the shear strength of unsaturated soil, φ' is the effective friction angle, c' is the effective cohesion. u_a is the pore air pressure, u_w is the pore water pressure, φ^b is the angle of shearing resistance with respect to matric suction. The single-phase calculation was conducted in the numerical simulation, which means the pore air pressure $u_a = 0$.

A series of experiments indicating that the angle of shearing resistance with respect to matric suction φ^b is not equal to the effective friction angle φ' and that φ^b

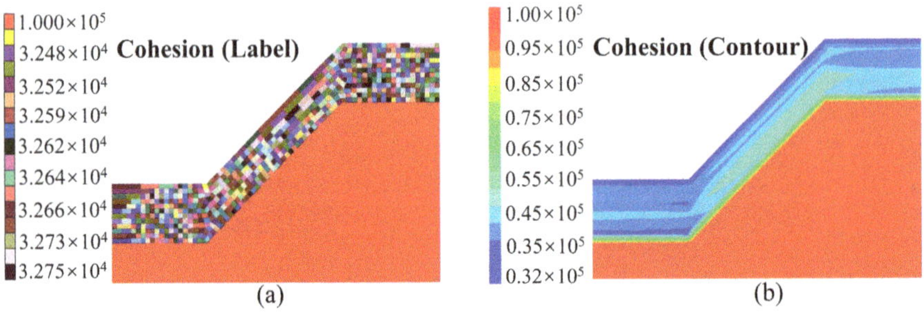

Fig. 4.26 The cohesion c of the slope before rainfall (**a**) Label, and (**b**) Contour (Pa)

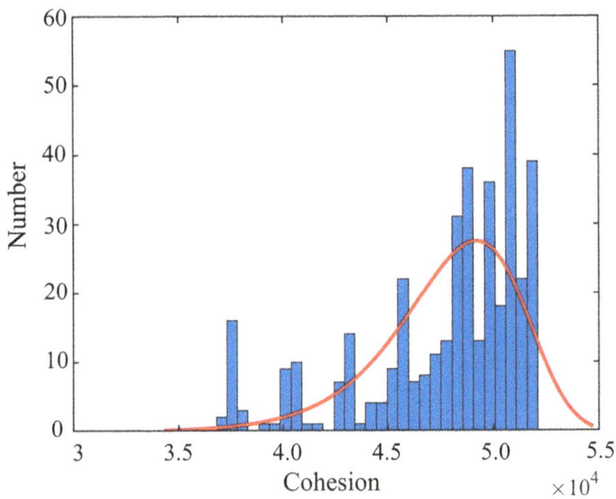

Fig. 4.27 The statistical data of modified cohesion (Pa) of moderately weathered layer in the slope

is not a constant value, but rather varies nonlinearly with suction (Nam et al. 2011). Many researchers still considered φ^b as a constant value, taking it as 15° based on empirical data (Ng and Shi 1998). This study also assumes a constant value for $\varphi^b = 15°$. The presence of matric suction is indicated when the pore pressure is less than 0 and the cohesion of that element needs to be modified. Otherwise, the shear strength parameters are left unchanged. Calculate the corrected cohesion c^* of the element considering the impact of matric suction according to the Eq. (4.10):

$$c^* = c' + \left(u_\mathrm{a} - u_\mathrm{w}\right)_\mathrm{f} \tan \varphi^\mathrm{b} \tag{4.10}$$

Figure 4.26 presents the results of the cohesion adjustment for the weathered layer of slope prior to rainfall, excluding the influence of matric suction. It is evident that the cohesion of the first weathered layer has significantly increased compared to the saturated state, which is characterized by a cohesion of 20 kPa. This indicates the critical role of matric suction in enhancing the initial shear strength of the weathered layer under unsaturated conditions. Similarly, Figure 4.27 shows the adjusted cohesion of the moderately weathered layer influenced by matric suction before the

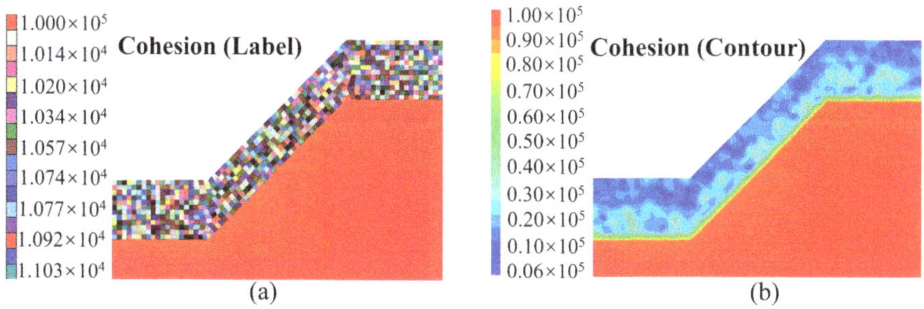

Fig. 4.28 (**a**) Weibull distribution of cohesion c (Pa). (**b**) Weibull distribution of cohesion c (Pa)

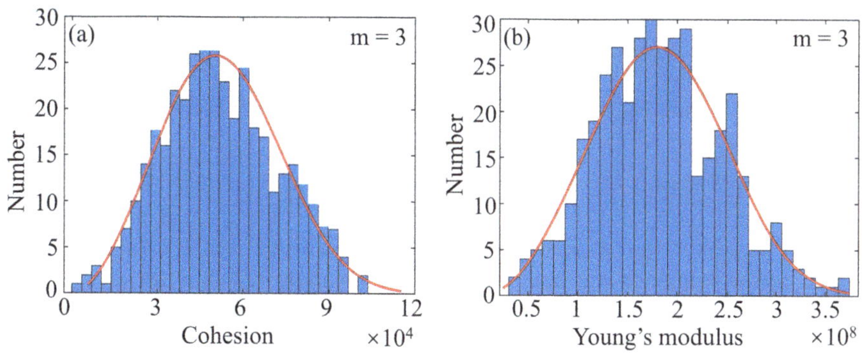

Fig. 4.29 The statistical data of cohesion c and Young's modulus E of Weibull distribution (Pa) with a degree of heterogeneity m=3

onset of rainfall. As rainfall progresses, the infiltration of rainwater leads to an increase in pore water pressure, transitioning from negative to positive values. This phenomenon, illustrated in Fig. 4.25, highlights the gradual dissipation of matric suction within the weathered layer of the slope. The reduction in matric suction corresponds to a decrease in the apparent cohesion of the soil, thereby diminishing the overall stability of the slope. This analysis underscores the dynamic relationship between matric suction, cohesion, and pore water pressure during rainfall events. The interplay of these factors significantly impacts the shear strength distribution within the slope, emphasizing the need for precise modeling of matric suction and its temporal changes to accurately predict slope stability under varying hydrological conditions.

Similar to the methodology described in Sect. 3.1, the heterogeneity of the weathered layers is modeled using a Weibull distribution to better represent the spatial variability of geotechnical properties. Figure 4.28 illustrates the Weibull distribution of cohesion c for the weathered layers within the slope, reflecting the inherent heterogeneity of the material. Both the cohesion and the elastic modulus for the strongly weathered layer were assumed to follow a Weibull distribution, with values ranging approximately between 0.25 to 1.75 times their respective mean values, as shown in Fig. 4.29. This wide range captures the stochastic nature of

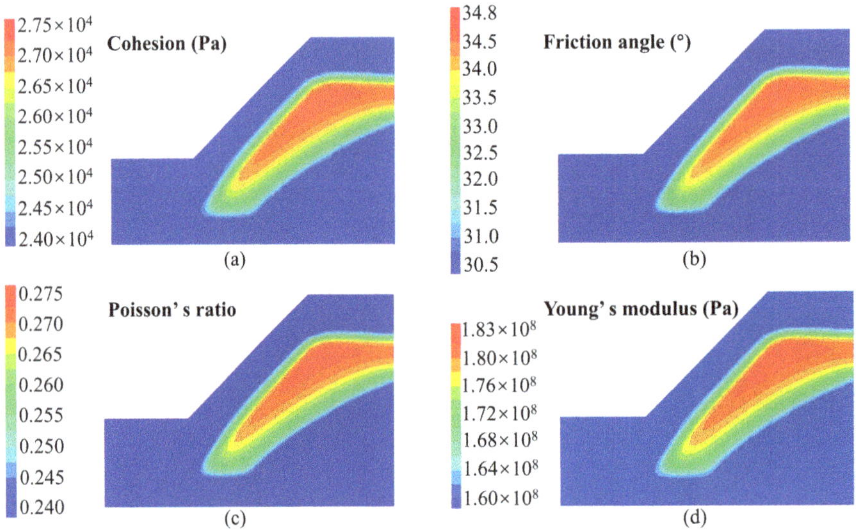

Fig. 4.30 Distributions of the mechanical parameters in the slope after 36 h of rainfall

material properties and their potential influence on slope stability. In contrast, a homogeneous approach was applied to the friction angle φ within the same weathered layer. This simplification assumes that the friction angle exhibits minimal variability compared to cohesion and elastic modulus, which aligns with observations from laboratory and field studies. By incorporating the Weibull distribution for key parameters, the analysis accounts for the non-uniform characteristics of weathered materials. This approach provides a more robust framework for evaluating slope stability under conditions influenced by rainfall infiltration. Future studies could extend this methodology to include spatial variability in friction angle φ and consider the correlation between different mechanical parameters to further refine the simulation accuracy.

To visually demonstrate the real-time variation of the rock mass mechanical parameters including cohesion c, internal friction angle φ, elastic modulus E, and Poisson's ratio μ with changes in saturation, the calculation sets the parameters for the strongly weathered layer listed in Table 4.6 as the parameters for the entire slope model. The resulting distribution of the mechanical parameters within the slope after 36 h of rainfall is shown in Fig. 4.30. It is evident that the distribution of these parameters is consistent with the saturation distribution after 36 h of rainfall within the slope, as shown in Fig. 4.25d. This calculation method, which considers the real-time changes in the properties of rocks, is more accurate than methods that only account for the dry and saturated states. By incorporating the gradual variations in material properties, such as cohesion and elastic modulus, within the weathered layers, the model offers a more realistic representation of rock behavior under rainfall conditions. As a result, the stability analysis becomes more reliable, effectively capturing the nuances of how rainfall infiltration influences stability of rock slopes.

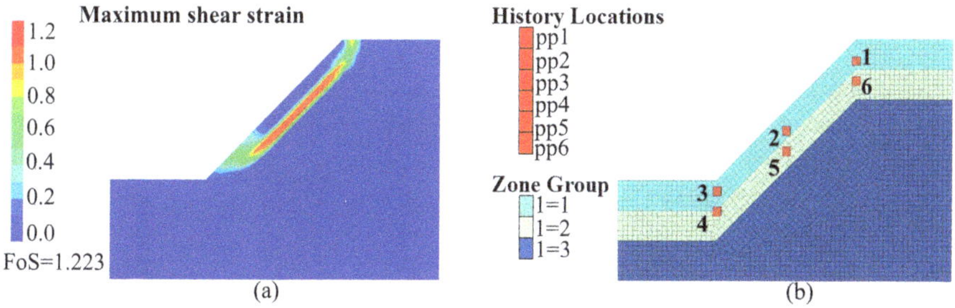

Fig. 4.31 (**a**) The maximum shear strain increment of slope after 48 h of rainfall (**b**) The position of monitoring points in the slope model

Fig. 4.32 Comparison of Factor of Safety (FoS) for heterogeneous slopes under real conditions (RC) and simplified conditions (SC)

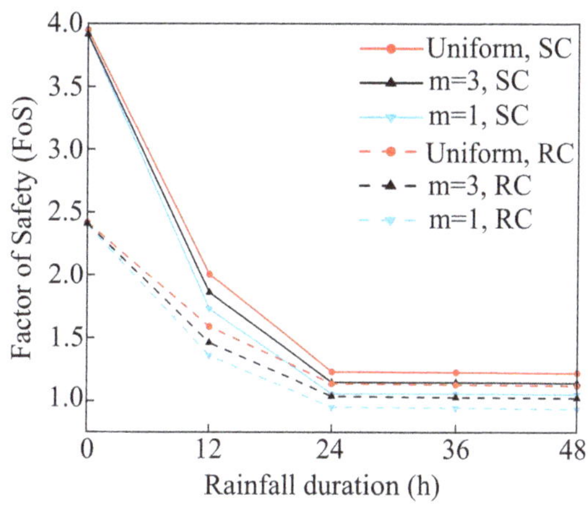

Based on the porosity of unsaturated areas, the reduction coefficient λ for rock strength parameters was determined. The real-time adjustment of the rock's physical and mechanical parameters can be achieved using FISH functions based on changes in saturation levels during rainfall infiltration. Figure 4.31a illustrates the maximum shear strain contour after 48 h of rainfall, revealing the formation of a distinct sliding zone in the shallow part of the slope. Figure 4.31b presents the six pore pressure monitoring points arranged within the slope. Same to the results in Sect. 4.1, the variation of pore pressure with rainfall duration is shown in Fig. 4.15. As shown in Fig. 4.32, the factor of safety (FoS) of the slope was significantly lower after 12 h of rainfall compared to the simplified scenario considering only dry and saturated conditions. This discrepancy arises because the large unsaturated area in the initial state results in a notable difference between the physical and mechanical parameters of the rock assigned under dry conditions and those adjusted to reflect actual saturation levels. After 24 h of rainfall, the difference in FoS between the two approaches diminishes as the saturated area in the slope rapidly expands, making the mechanical parameters of rocks approach saturation conditions. Besides, considering rock mass heterogeneity further reduces the FoS compared to a

Table 4.7 The calculation results of two-dimensional slope model

Rainfall duration (h)	FoS of the slope after rainfall					
	Only dry and saturated conditions			Real saturation conditions		
	Homogeneous	$m = 1$	$m = 3$	Homogeneous	$m = 1$	$m = 3$
0	3.95	3.91	3.914	2.42	2.40	2.41
12	2.008	1.73	1.86	1.590	1.360	1.461
24	1.230	1.055	1.145	1.137	0.949	1.035
36	1.227	1.055	1.145	1.129	0.943	1.027
48	1.22	1.051	1.14	1.121	0.934	1.02

homogeneous slope after the same duration of rainfall. Notably, for the most heterogeneous state of the rock mass with $m = 1$, the FoS is 0.949 after 24 h of rainfall. This indicates the slope failure that aligns with the actual condition observed in the site. In contrast, the FoS for a homogeneous slope is 1.137, suggesting the stable condition of slope after continuous 24 h of heavy rainfall. Table 4.7 summarizes the calculation results of two-dimensional slope model under two different conditions. This illustrates the significant impact that accounting for heterogeneity has on slope safety assessments under short-term heavy rainfall.

4.3 Slope Surface Undulation of 2D Slopes

This section primarily examines the influence of slope surface undulation and angle on slope stability under rainfall conditions using 2D slope models in FLAC^{3D}. It also briefly compares the factors of safety (FoS) obtained from 2D analyses with those from 3D analyses. As illustrated in Fig. 4.33, the 3D slope model consists of two weathered layers, each 3 m thick, representing strongly weathered and moderately weathered rocks. To isolate the impact of slope surface geometry on slope stability, 2D cross-sections were extracted by rotating clockwise around the AB axis, creating a cross-section every 5°, as shown in Fig. 4.33b. The numbering of all 19 cross-sections is displayed in Fig. 4.34. The numerical simulations used the mechanical parameters for a saturated slope, as specified in Table 3.3. Figures 4.35 and 4.36 present schematic diagrams of cross-sections No. 1 to 6 and No. 7 to 12, respectively. Due to minimal geometric differences, cross-sections No. 13 to 19 are excluded from detailed analysis. A comparative analysis identified representative cross-sections: cross-section 3 for sections No. 1–5, cross-sections 6 and 9 for sections No. 6–10, and cross-section 12 for sections No. 11–19. Figures 4.37, 4.38, 4.39 and 4.40 illustrate the selected cross-sections 3, 6, 9, and 12, with AB representing the groundwater level. The slope height for all four cross-sections is 42 m, ensuring that differences in FoS arise primarily from variations in slope surface geometry. The assumptions regarding matric suction remain consistent with those introduced in the previous section. Figure 4.41 shows the distribution of saturation across the slopes of the four cross-sections after 24 h of rainfall, while Fig. 4.42

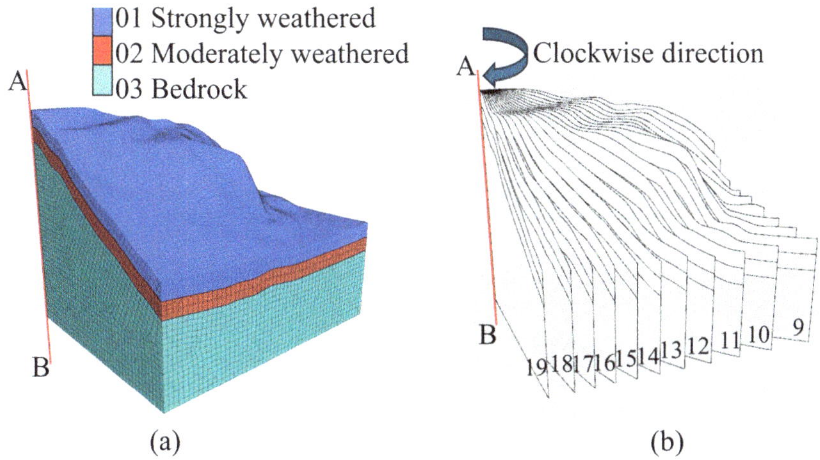

Fig. 4.33 (**a**) 3D slope model in FLAC^{3D}; (**b**) Clockwise direction for extracting 2D cross-sections

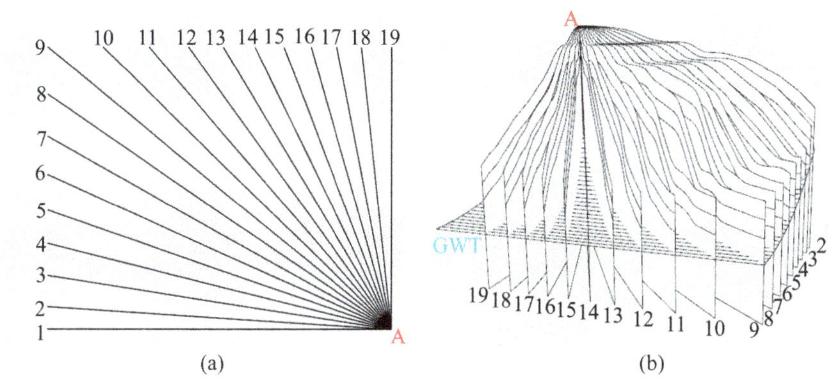

Fig. 4.34 Numbering of 2D cross-sections: (**a**) Plan view; (**b**) Elevation view

depicts the contours of maximum shear strain and the corresponding FoS for each cross-section. Results indicate that the FoS values obtained from 2D slope models after 24 h of rainfall are consistently lower than those derived from 3D analyses under identical conditions. Using 2D cross-sections 3, 9, and 12 as examples, Figure 4.43a outlines a method for simplifying the slope surface geometry. If the slope surface undulation height does not exceed 2 m and the undulation angle is less than 20° relative to the baseline, the slope surface can be simplified to a straight line. Figure 4.43b–d display the FoS values for the simplified cross-sections 3.1, 9.1, and 12.1 after 24 h of rainfall. The results show minimal differences between the FoS of the simplified and original cross-sections (refer to Fig. 4.42). Therefore, when slope surface undulations are limited to a height of 2 m and an angle of 20° or less, the

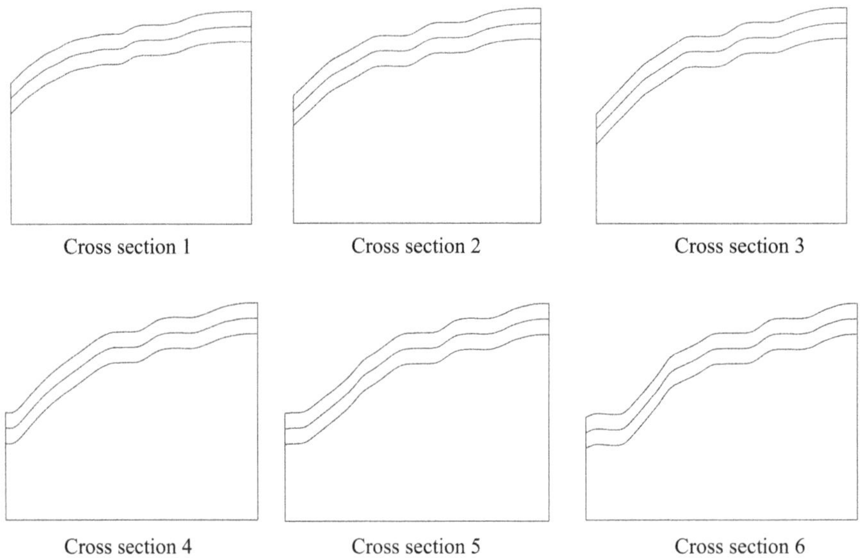

Fig. 4.35 Cross sections 1 to 6

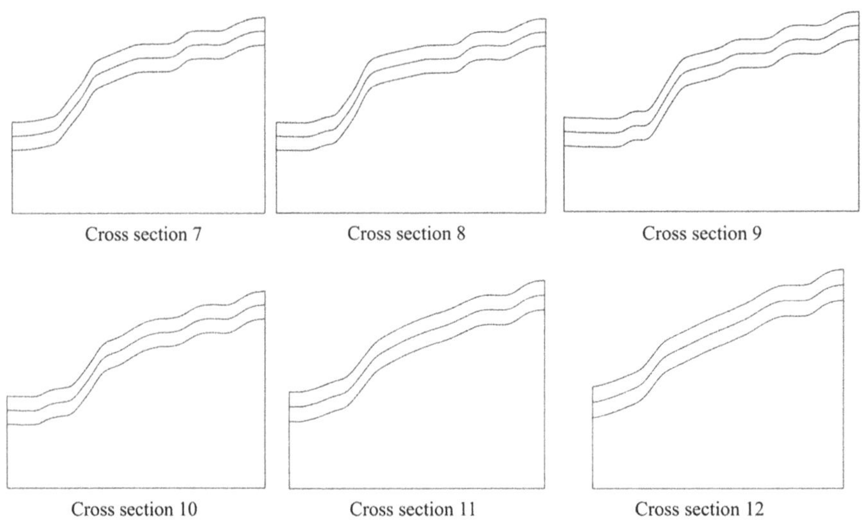

Fig. 4.36 Cross sections 7 to 12

calculated FoS closely approximates that of a slope treated with a straight-line surface. For the 3D slope model used in this study (Fig. 4.33), when the 2D slope surface exhibits minor undulations and insignificant angles of variation, the factor of safety (FoS) after 24 h of rainfall is comparable to that of a simplified straight-line slope. A more detailed investigation into the effects of slope surface undulation

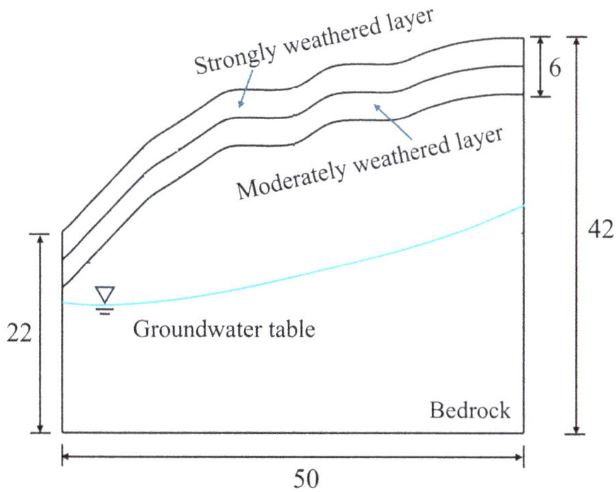

Fig. 4.37 2D Cross-Section No. 3 (unit: m)

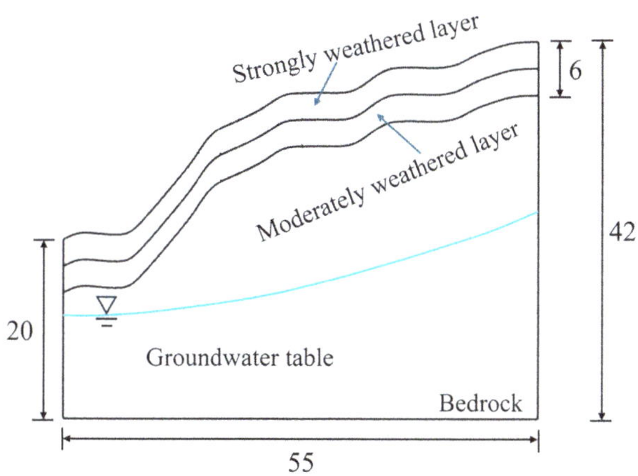

Fig. 4.38 2D Cross-Section No. 6 (unit: m)

height and angle on slope stability under rainfall conditions will be conducted in future research.

Table 4.8 presents the differences in factor of safety (DF_1) between the 2D cross-sections and the 3D slope model in the stability analysis ((F_{3D}—F_{2D})/F_{3D}, or $\Delta F_S/F_{3D}$). Results indicate that the factors of safety (FoS) for all 2D cross-sections are lower than those for the 3D slope after 12, 24, 36, and 48 h of rainfall, except for cross-section 12. This is because, when assuming the groundwater level of the 3D slope as a single plane, the method of rotating and cutting 2D cross-sections results in differences in the groundwater level for each 2D cross-section, as shown in Fig. 4.34b. Clearly, the groundwater levels for cross-sections 9 to 19 are lower than those for cross-sections 1 to 8. Consequently, the factor of safety for cross-section

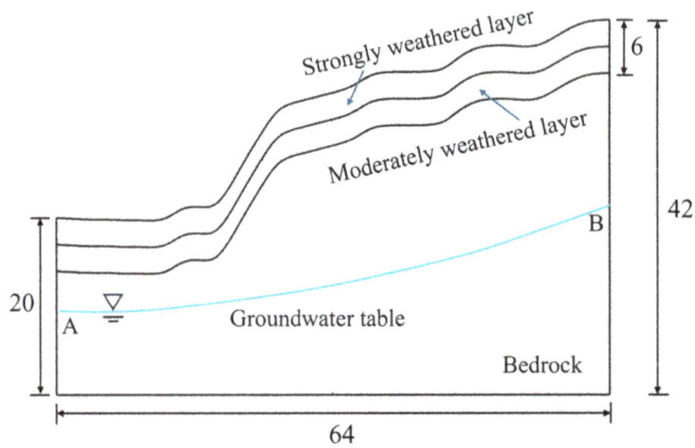

Fig. 4.39 2D Cross-Section No. 9 (unit: m)

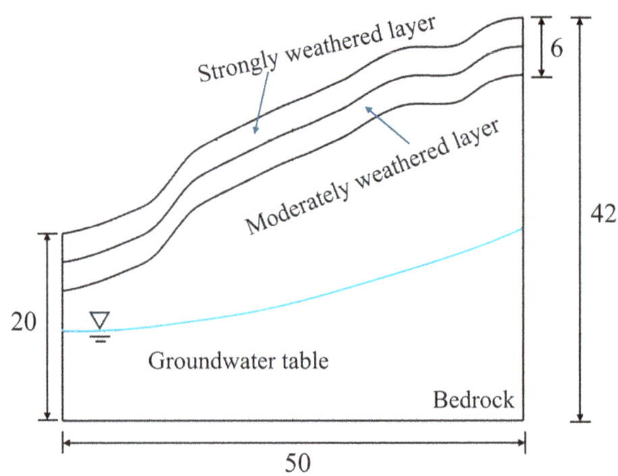

Fig. 4.40 2D Cross-Section No. 12 (unit: m)

12 is higher than that of the 3D slope. Table 4.8 illustrates a comparison of the factor of safety (FoS) over time between the 3D slope model and the 2D cross-sections. These comparisons provide insights into how surface geometry and groundwater table influence slope stability assessments under heavy rainfall. Throughout the 48 h of rainfall, the safety factor of cross section 9 consistently fell below that of the 3D slope, while that of cross section 12 consistently exceeded it. This observation suggests that relying on selected cross-sections does not always yield more conservative safety factor estimates compared to 3D slope stability analyses, thereby challenging the traditional assumption commonly held in geotechnical engineering. Specifically, the factor of safety for cross-section 12 exceeds that of the three-dimensional slope, reaching 1.31 after 48 h of rainfall. This reflects a relatively stable condition in contrast to cross-section 9, which approaches failure with a factor of safety of only 1.039. These results highlight that variations in surface geometry and groundwater conditions of 2D slopes can significantly affect the calculated

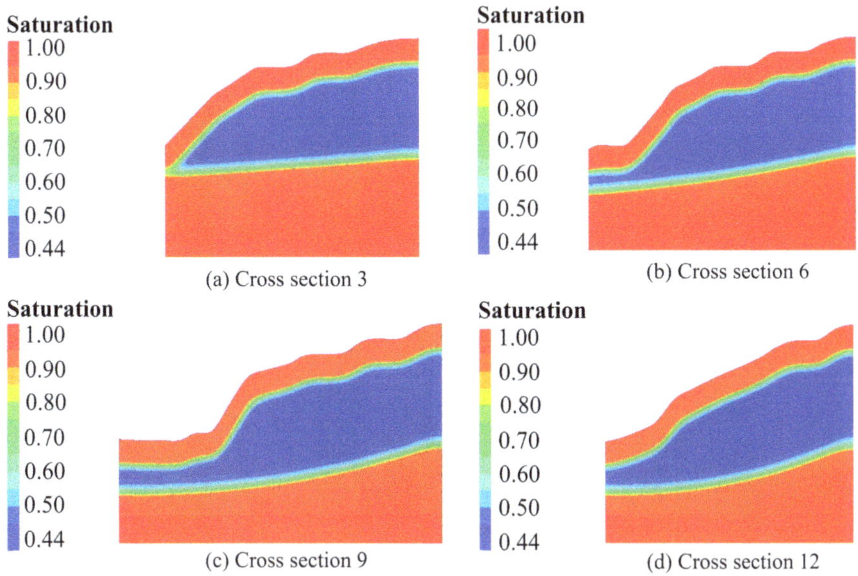

Fig. 4.41 Distribution of saturation in the slope after 24 h of rainfall. (**a**) Cross section 3; (**b**) Cross section 6; (**c**) Cross section 9; (**d**) Cross section 12

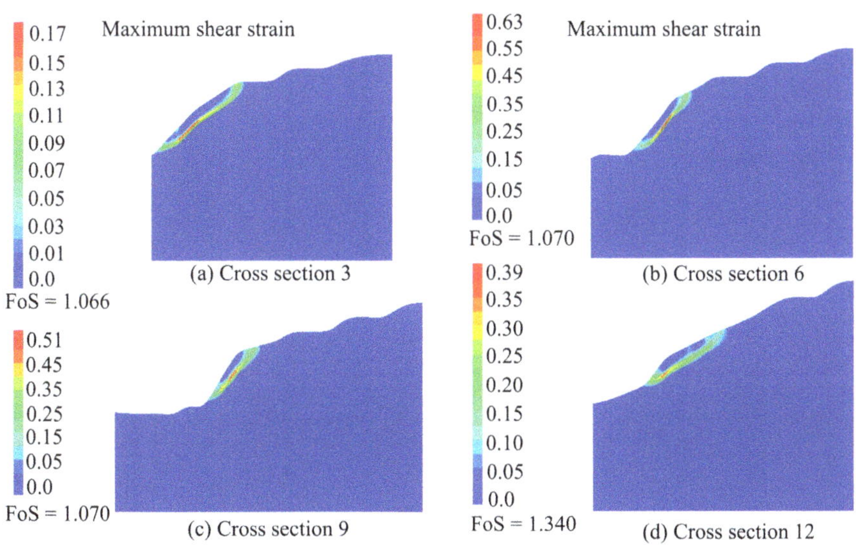

Fig. 4.42 Contour of maximum shear strain in the slope after 24 h of rainfall. (**a**) Cross section 3; (**b**) Cross section 6; (**c**) Cross section 9; (**d**) Cross section 12

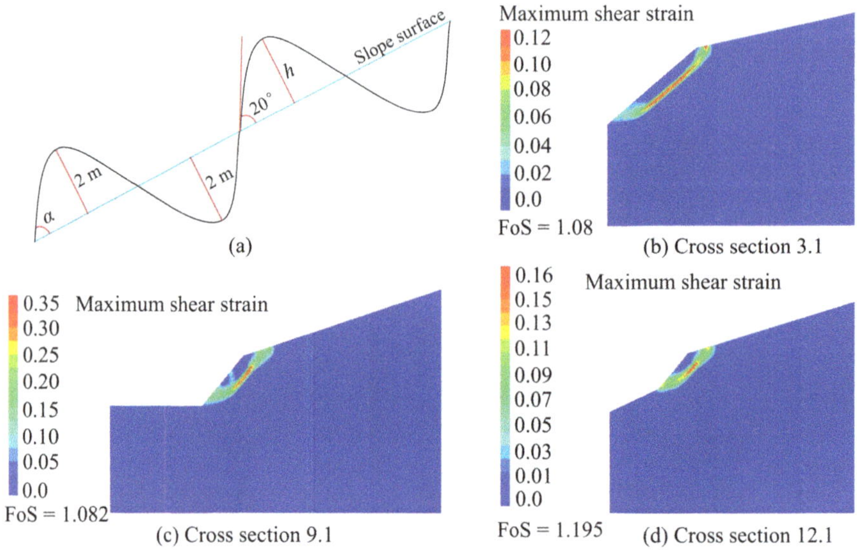

Fig. 4.43 (**a**) Schematic diagram of undulation degree and angle of slope surface; Contour of maximum shear strain in the slope after 24 h of rainfall. (**b**) Cross section 3.1; (**c**) Cross section 9.1; (**d**) Cross section 12.1

Table 4.8 The calculation results of the factor of safety (FoS) for 3D and 2D slopes

Models		Rainfall duration (h)				
		0	12	24	36	48
3D slope		3.52	2.164	1.238	1.191	1.152
2D cross sections	No.3	3.398	1.793	1.066	1.055	1.043
	DF_1 (%)	3.5	17.1	13.9	11.4	9.5
	DF_2 (%)	–	18.7	1.3	1.4	1.1
	No.6	3.34	1.789	1.07	1.06	1.059
	DF_1 (%)	5.1	17.3	13.6	10.9	8.1
	No.9	3.367	1.772	1.07	1.051	1.039
	DF_1 (%)	4.3	18.1	13.6	11.8	9.8
	DF_2 (%)	–	23.9	1.1	2.7	2.5
	No.12	3.73	2.32	1.340	1.332	1.31
	DF_1(%)	−6.0	−7.2	−8.2	−11.8	−13.7
	DF_2 (%)	–	17.1	−8.2	−7.9	−7.8

factor of safety. These variations arise from the selection of different cross-sections. This influence becomes particularly pronounced under rainfall conditions. In reality, the groundwater level of a 3D slope is much more complex, which significantly increases the likelihood of calculation inaccuracies when selecting 2D cross-sections due to variations in the groundwater level. Therefore, it is strongly recommended that engineers perform comparative calculations for both 2D and 3D slopes to enhance the safety of engineering designs.

Table 4.8 also presents the differences in factors of safety (DF_2) between the real slope surface and the simplified slope surface in the slope stability analysis ((F_{SD}—F_{2D})/F_{SD}, or $\Delta F_S/F_{SD}$). The difference in safety factors between cross sections 9 and 9.1 is similar to that observed between cross section 6 and 6.1, and it is not presented in the table. Comparing the actual 2D cross-sections with the simplified 2D cross-sections, significant differences in FoS are observed after 12 h of rainfall, particularly for cross-section 9 and its simplified counterpart (cross-section 9.1), where the FoS difference reaches nearly 24%. However, this difference decreases rapidly as rainfall duration increases. After 24 and 48 h of rainfall, the differences in FoS between the actual and simplified 2D cross-sections become negligible for all three groups of cross-sections. The reduction in the difference between the actual and simplified 2D cross-sections with increasing rainfall duration can be attributed to the gradual saturation of the slope. As the rainfall continues, the influence of the slope surface geometry becomes less significant because the entire slope, including the undulated sections, reaches a state of near-saturation. In this saturated condition, the overall stability of the slope is more dependent on the material properties and less on the geometric variations of the slope surface. Consequently, the factor of safety for both the actual and simplified cross-sections converges as the slope becomes more uniformly saturated, minimizing the effect of surface undulations. The significant difference in the factor of safety (FoS) before 12 h of rainfall can be attributed to the matric suction. During the initial stages of rainfall, the matric suction plays a crucial role in the stability of the slope. The unsaturated zone above the groundwater table has varying matric suction values, which affect the shear strength of the soil. For the actual 2D cross-sections, the geometry of the slope influences how water infiltrates and affects the distribution of matric suction. In contrast, the simplified 2D cross-sections assume a uniform or simplified geometry, which may not accurately represent the variation in matric suction, leading to a larger difference in FoS before 12 h of rainfall. A significant decrease in the factor of safety between 12 and 24 h of rainfall was observed in Table 4.8. This decline is primarily due to the rapid increase in pore water pressure within the shallow layers of the slope, which reduces the effective stress and consequently lowers the shear strength of soil and rock mass. After 24 h of rainfall, the change in the safety factor becomes less significant, as the large difference in permeability coefficients between the weathered layer and the bedrock slows the infiltration rate of rainwater into the bedrock. As a result, the pore water pressure in the bedrock does not change significantly and the change in the safety factor of the slope becomes less significant. This demonstrates the significant impact of the permeability coefficient of the soil and rock mass on slope stability under rainfall conditions.

4.4 Conclusions

This chapter begins by implementing the theoretical framework presented in the previous chapter using a standard slope model. It provides a detailed demonstration of the effects of applying a Weibull distribution to the weathered layer of the slope. The chapter compares the variations in safety factors of the slope under rainfall conditions caused by different degrees of heterogeneity in the weathered layer. The Weibull distribution has proven to be an effective and relatively accurate method for describing the heterogeneity of weathered layers in slopes. A higher value of m corresponds to a greater safety factor for the slope, indicating that the less uniform the slope, the lower its stability under rainfall conditions. Incorporating an appropriate heterogeneity index m is crucial for accurately assessing slope stability. This chapter also concludes that the relationship between the permeability coefficient of the slope's weathered layers and rainfall intensity has a significant impact on slope stability. When the saturated permeability coefficient of the weathered layer is slightly greater than the rainfall intensity, the safety factor of the slope decreases most rapidly within the same rainfall duration. This finding underscores the importance of accurately characterizing the permeability properties of the weathered layer to better predict slope stability under varying rainfall conditions. Furthermore, this chapter simulates the peak strengths of soft rocks with varying degrees of homogeneity and porosity under uniaxial compression test in FLAC^{3D}. These simulations reveal the influence of heterogeneity and porosity on the mechanical behavior of rocks, providing insights into their impact on slope stability. The results demonstrate how these factors interact under rainfall infiltration, contributing to a deeper understanding of the mechanisms driving changes in slope stability during heavy rainfall. Rocks with higher porosity generally exhibit more pronounced water-induced weakening effects. Uniaxial compression tests have shown that under identical saturation and heterogeneity conditions, the peak strength of sandstone samples decreases as porosity increases. For sandstone samples under the same saturation conditions and with the same degree of heterogeneity, peak strength σ_{peak} decreases as porosity increases. Similarly, with the same porosity and saturation conditions, higher heterogeneity of mechanical parameters results in lower peak strength of sandstone samples. For rock samples with identical porosity and homogeneity, greater water content also reduces peak strength. In the 2D slope model simulation, the real-time adjustment of physical and mechanical parameters based on saturation levels provides a more accurate and realistic representation of rock behavior under rainfall conditions, compared to simplified methods that differentiate only between dry and saturated states. This analysis highlights the importance of considering the varying strengths of rock masses in saturated conditions, particularly due to differences in porosity across weathered layers. This study also demonstrates that the dissipation of matric suction within weathered layers is a critical factor in reducing slope stability during rainfall. As rainfall progresses, the unsaturated zone within the slope gradually diminishes, accompanied by the dissipation of matric suction. This process leads to a significant reduction in cohesion c and elastic modulus E

within the weathered layers, ultimately causing a decline in slope stability under short-term heavy rainfall. The results underscore the significant impact of weathered layer heterogeneity on slope stability. Slope safety factors obtained using a heterogeneous model ($m = 1$) and real-time parameter adjustments based on saturation (2D FoS = 0.934, 3D FoS = 1.06) closely align with observed slope failure conditions. In contrast, results from the homogeneous model (2D FoS = 1.12, 3D FoS = 1.191) are overly conservative and fail to reflect actual slope behavior. This highlights the importance of accounting for heterogeneity and dynamic adjustments in slope stability analyses under rainfall conditions. Finally, the chapter briefly explores the influence of the geometric shape of a two-dimensional slope surface on the safety factor under rainfall conditions. When the undulations of the slope surface are minor (elevation differences not exceeding 5 m and undulation angles not exceeding 20°), the calculated safety factor for a simplified straight-line slope tends to converge with that of the actual slope as rainfall duration increases. This finding highlights the feasibility of using simplified slope geometries for certain rainfall infiltration scenarios without significantly compromising accuracy.

References

Cai F, Ugai K (2004) Numerical analysis of rainfall effects on slope stability. Int J Geomech 4:69–78. https://doi.org/10.1061/(ASCE)1532-3641(2004)4:2(69)

Chen X, Zhang L, Zhang L et al (2021) Modelling rainfall-induced landslides from initiation of instability to post-failure. Comput Geotech 129:103877. https://doi.org/10.1016/j.compgeo.2020.103877

Cho SE, Lee SR (2001) Instability of unsaturated soil slopes due to infiltration. Comput Geotech 28:185–208. https://doi.org/10.1016/S0266-352X(00)00027-6

Cho MTT, Chueasamat A, Hori T et al (2021) Effectiveness of filter gabions against slope failure due to heavy rainfall. Soils Found 61:480–495. https://doi.org/10.1016/j.sandf.2021.01.010

Cundall P, Board M (1988) A microcomputer program for modelling large-strain plasticity problems. In: Proceedings of the Sixth International Conference on Numerical Methods in Geomechanics, Innsbruck, Austria

Fredlund DG, Morgenstern NR, Widger RA (1978) The shear strength of unsaturated soils. Can Geotech J 15:313–321. https://doi.org/10.1139/t78-029

Fu J, Haeri H, Sarfarazi V et al (2022) Extended finite element method simulation and experimental test on failure behavior of defects under uniaxial compression. Mech Adv Mater Struct 29:6966–6981. https://doi.org/10.1080/15376494.2021.1989730

Hoyos LR, Suescún-Florez EA, Puppala AJ (2015) Stiffness of intermediate unsaturated soil from simultaneous suction-controlled resonant column and bender element testing. Eng Geol 188:10–28. https://doi.org/10.1016/j.enggeo.2015.01.014

Iyare UC, Blake OO, Frash LP, et al (2023) Water-Weakening Effects on the Failure Behavior of Mudstones. Rock Mech Rock Eng 56:9171–9185. https://doi.org/10.1007/s00603-023-03544-4

Li X, Jiang Y, Sugimoto S (2024) Slope stability analysis under heavy rainfall considering the heterogeneity of weathered layers. Landslides. https://doi.org/10.1007/s10346-024-02404-8

Li X, Jiang Y, Sugimoto S (2025) Slope stability analysis under heavy rainfall considering the heterogeneity of weathered layers. Landslides 22:1257–1273. https://doi.org/10.1007/s10346-024-02404-8

Nam S, Gutierrez M, Diplas P, Petrie J (2011) Determination of the shear strength of unsaturated soils using the multistage direct shear test. Eng Geol 122:272–280. https://doi.org/10.1016/j.enggeo.2011.06.003
Ng CWW, Shi Q (1998) A numerical investigation of the stability of unsaturated soil slopes subjected to transient seepage. Comput Geotech 22:1–28. https://doi.org/10.1016/S0266-352X(97)00036-0
van Genuchten MT (1980) A closed-form equation for predicting the hydraulic conductivity of unsaturated soils. Soil Sci Soc Am J 44:892–898. https://doi.org/10.2136/sssaj1980.03615995004400050002x
Pan Y, Wu G, Zhao Z, He L (2020) Analysis of rock slope stability under rainfall conditions considering the water-induced weakening of rock. Computers and Geotechnics 128:103806. https://doi.org/10.1016/j.compgeo.2020.103806
Wang J, Zhang D, Wang N, Gu T (2019) Mechanisms of wetting-induced loess slope failures. Landslides 16:937–953. https://doi.org/10.1007/s10346-019-01144-4
Wang D, Wang B, Yuan W, Liu L (2023) Investigation of rainfall intensity on the slope failure process using GPU-accelerated coupled MPM. Comput Geotech 163:105718. https://doi.org/10.1016/j.compgeo.2023.105718
Xie W-Q, Liu X-L, Zhang X-P et al (2024) A review of test methods for uniaxial compressive strength of rocks: theory, apparatus and data processing. J Rock Mech Geotech Eng 17:1889–1905. https://doi.org/10.1016/j.jrmge.2024.05.003
Yarali O, Kahraman S (2011) The drillability assessment of rocks using the different brittleness values. Tunn Undergr Space Technol 26:406–414. https://doi.org/10.1016/j.tust.2010.11.013
Zhang LL, Fredlund MD, Fredlund DG et al (2015) The influence of the unsaturated soil zone on 2-D and 3-D slope stability analyses. Eng Geol 193:374–383. https://doi.org/10.1016/j.enggeo.2015.05.011

Chapter 5
Case Study of Rainfall-Induced Landslides

Abstract This chapter presents a detailed analysis of the Sasebo landslide, which occurred on September 17, 2023, in Kosaza Town, Sasebo City, Nagasaki, Japan. The study builds upon the theoretical methods and governing equations introduced in previous chapters. It incorporates field borehole data to investigate the mechanisms of slope failure. The results show that considering the heterogeneity of weathered soils significantly improves the accuracy of numerical simulations. A Weibull distribution is applied to represent the variability in soil cohesion c and elastic modulus E. This approach leads to numerical simulations that closely replicate the actual failure patterns observed at the site. In addition, this chapter compares the results of both two-dimensional (2D) cross-sectional and three-dimensional (3D) slope stability analyses. The comparison highlights the significant influence of spatial variability within the weathered layer on slope behavior. These findings reinforce the necessity of accounting for soil heterogeneity in slope stability assessments under rainfall conditions.

Keywords Case study · Sasebo landslide · 2D versus 3D · Heterogeneity · Weathered layers

5.1 Sasebo Landslide (2D)

The Sasebo landslide is located in Kosaza Town, Sasebo City, Nagasaki, Japan (Fig. 5.1a). A zoomed-in view of the site shows three distinct sliding zones identified during the field investigation (Fig. 5.1b). The affected slope faces north–northeast with an inclination of approximately 21°. The landslide measures 109 m in length and 89 m in width, with an average thickness of 4.3 m (Fig. 5.1c). The average slope height is approximately 20 m. The landslide covers an area of approximately 5700 m^2, with an estimated volume of nearly 90,000 m^3. Sliding Zone 1, located on the left side, is estimated to be 56 m wide and 71 m long, whereas Sliding

Y. Jiang, X. Li, *Stability Evaluation of Slope Under Heavy Rainfall Based on Geometric Features*, Springer Natural Hazards,
https://doi.org/10.1007/978-981-95-8656-1_5

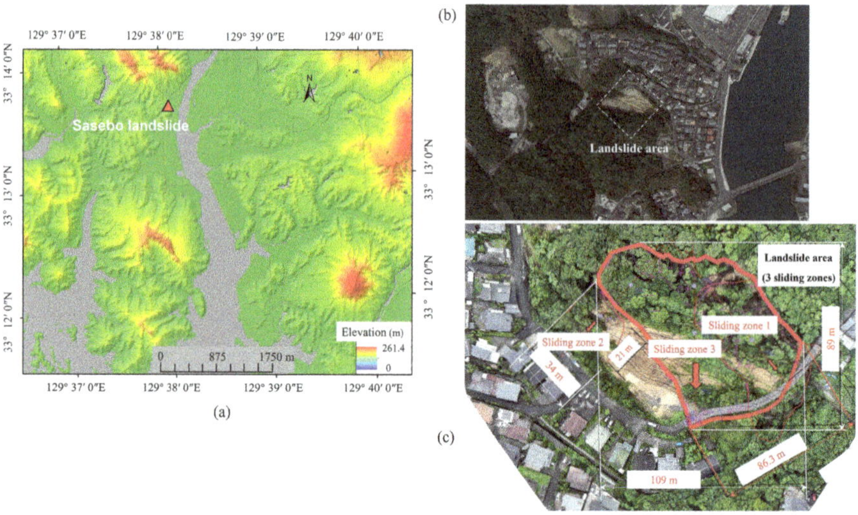

Fig. 5.1 (**a**) The location of Sasebo landslide (33°13′36.6″N, 129°38′4.3″E). (**b**) A zoomed-in view of the Sasebo landslide. (**c**) Full view of the three sliding zones (Li et al. 2025)

Fig. 5.2 Detailed photos of the Sasebo landslide: (**a**) Head scarp, (**b**) Step-like cracks, (**c**) Retaining wall deformation, (**d**) Compression cracks, (**e**) Step-like cracks, (**f**) Cracks at slope heads

Zone 2 on the right is about 34 m wide and 21 m long. Sliding Zone 3 exhibits the most significant deformation and represents the main body of the landslide. At the head scarp, continuous east–west–oriented sliding scars are visible along the ridge (Fig. 5.2a). Subsidence-related cracks and step-like features are evident downslope (Fig. 5.2b). At the toe, extrusion deformation is observed against the gravity retaining wall (Fig. 5.2c). On the northern side of the slope, compression cracks appear

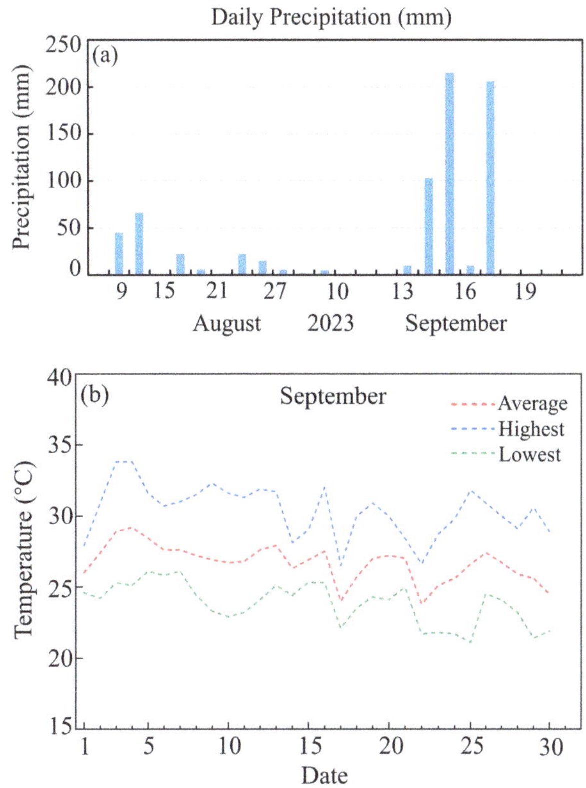

Fig. 5.3 (**a**) Daily precipitation and (**b**) temperature for September in the landslide area (Li et al. 2025)

on the road surface (Fig. 5.2d). On the left flank, step-like cracks extending downslope in the north–south direction are present (Fig. 5.2e), accompanied by subsidence features similar to those at the head. The right flank contains intermittent north–south oriented cracks extending from the head (Fig. 5.2f). From September 13 to 17, 2023, Sasebo City recorded 544 mm of rainfall, including a maximum daily rainfall of 215 mm and a maximum hourly rainfall of 59 mm (Fig. 5.3a). This extreme rainfall event triggered the landslide, resulting in the collapse of a city road (Tahara–Kuroishi Line) at 9:00 a.m. on September 17.

The mechanism of slope failure induced by short-duration heavy rainfall depends strongly on the initial hydrological conditions and the soil–water characteristic curves (SWCC), which govern unsaturated soil behavior (Ng and Shi 1998; Cai and Ugai 2004). Accordingly, a detailed geological investigation of the Sasebo landslide was carried out to support the analysis. Six boreholes were drilled to obtain core samples and monitor groundwater levels (Fig. 5.4). As shown in Fig. 5.3a, rainfall in September was concentrated between September 13 and 17. Soil samples collected from the landslide toe in late September had degrees of saturation ranging from 0.29 to 0.33, which were adopted as the initial saturation in the numerical simulation. Based on laboratory SWCC tests, matric suction from the ground surface to a depth of 5 m was estimated to range from 150 to 100 kPa, with a residual

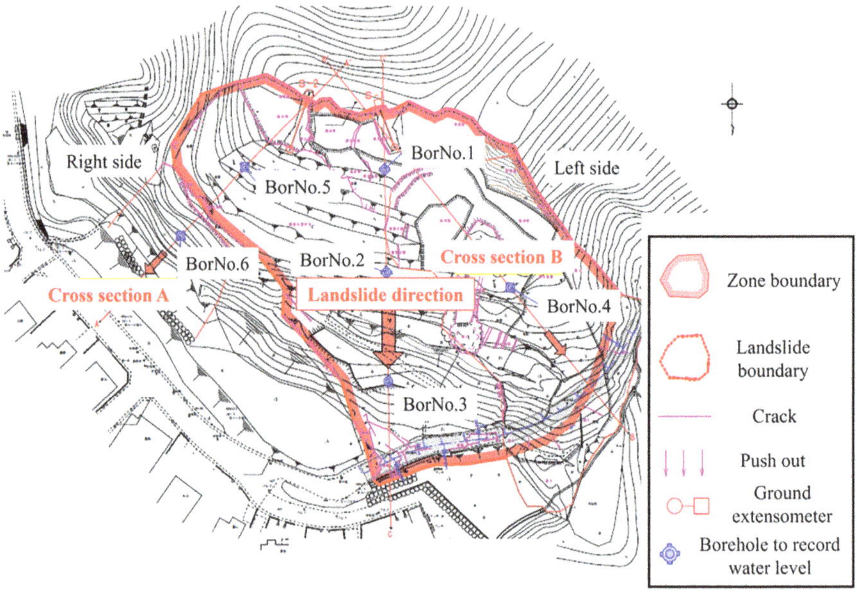

Fig. 5.4 Positions of six boreholes and three representative cross sections (Li et al. 2025)

saturation of $S_r = 0.065$ (Li et al. 2025). The slope experienced almost no rainfall from August 27 to September 12, prior to the critical event. Combined with high ambient temperatures during August and September (Fig. 5.3b) and the relatively high permeability of the slope soils (6×10^{-6} m/s), the shallow slope materials likely underwent significant drying due to evaporation and drainage. Only light rainfall occurred before the major event, and the two-week dry period prior to September 13 was comparable to the dry period following the event (September 17–30). Therefore, using the late-September sampling data to represent the initial moisture condition prior to the critical rainfall event is considered reasonable.

Borehole data collected after the landslide indicate that the groundwater level was slightly above the sliding surface. It is therefore hypothesized that this elevated groundwater level reduced the shear strength of a weak clay interlayer, initiating sliding along the clay layer at a depth of approximately 10 m (Fig. 5.6). This clay layer, about 2 cm thick and observed in 5 of the 6 boreholes, exhibits a flow-like morphology relative to the slope. This suggests that it acted as the primary sliding surface. The initial movement along the deeper clay layer likely triggered the subsequent sliding of the overlying weathered soil at a depth of 5 m. It was driven by a similar reduction in shear strength due to rainfall infiltration. However, the actual slope failure was confined to the upper 5 m of the slope. At 10 m depth, the slope exhibits slow, continuous movement along the slip surface without experiencing full failure. As shown in Fig. 5.5, Block ABCD remains intact and only exhibits gradual displacement along the estimated slip surface 2. The identification of this 10 m slip surface is supported by monitoring data from six boreholes. To verify this potential

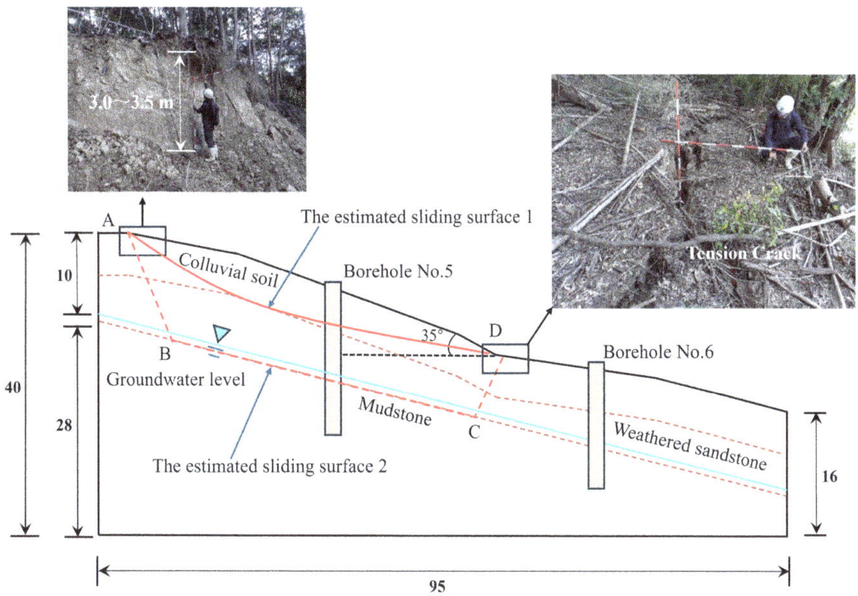

Fig. 5.5 Geological cross-section A of the Sasebo landslide (unit: m) (Li et al. 2025)

Table 5.1 Deformation monitoring data of the slope at a depth of 9 ~ 11 m on March 29, 2024

Boreholes	No. 1	No. 2	No. 3	No. 4	No. 5
Depth (m)	11	10	9	10	10
Deformation (μ strain)	207	27	0	151	199

slip surface, strain gauges have been continuously monitored in Boreholes No. 1–6 since December 14, 2023. Pipe-type extensometers were installed in Boreholes No. 1–5 along the slide direction indicated in Fig. 5.4 to monitor deformations at depths of 10–11 m. Monitoring results indicate that the slope continues to move slowly along slip surface 2, with μ strain representing the strain measured by the extensometers (1 μ strain = 1 × 10^{-6} m). For instance, in Borehole No. 4, cumulative deformation at 10 m depth reached 151 μ over 106 days, while at 11 m depth, it reached 161 μ (Fig. 5.7). Here, cumulative deformation represents the total displacement of the sliding mass over the monitoring period. In contrast, Borehole No. 3 recorded no detectable movement at 9 m depth near the estimated slip surface. Table 5.1 summarizes the deformation monitoring data for depths of 9–11 m as of March 29, 2024.

Cross-section A was selected for analysis and the average slope gradient of the cross-section is 33° (Fig. 5.5). As shown in Fig. 5.6, there is a mixture of colluvial soil and strongly weathered sandstone from 0 m to 5 m in the slope. From depth of 5 m to 10 m, the layer mainly contains moderately weathered sandstone. The fresh sandstone is distributed below the depth of 10 m. Table 5.2 summarizes the geological characterization of the Sasebo landslide. Based on the geotechnical investigation

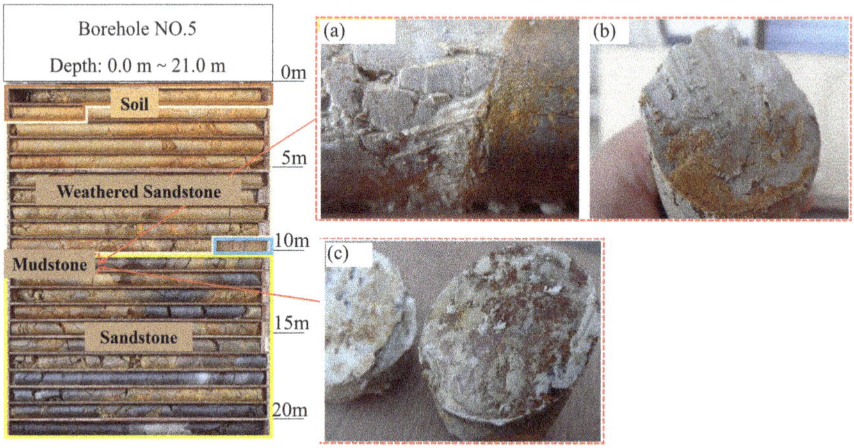

Fig. 5.6 Soil samples from Borehole NO. 5. (**a**) The estimated slide surface 2 in the depth of 10 m: a layer of white clay in the interface between mudstone and weathered sandstone. Obvious (**b**) scratch marks and (**c**) polished surface were observed at the interface (Li et al. 2025)

Table 5.2 The geological data of the Sasebo landslide

Stratum		Designation	Classification	Description
Quaternary	Colluvial soil	Dt	Sandy soil	Coarse-grained soil with sub-angular gravel of sandstone. The matrix is predominantly fine-grained sand with gravel size mainly ranges from 10 to 30 mm
Neogene	Weathered sandstone	WSs	Weathered rock	The hardness of weathered rock allows it to be broken with light hammer strikes
	Weathered mudstone	Ms	Soft rock	Fragile mudstone with weathered sandstone. Some parts of the core are so weakened that can be crumbled by hand
	Sandstone	Ss	Soft rock	Fresh fine-grained sandstone. The hardness of rock allows it to be broken under medium hammer strikes

report, the geotechnical properties of the slope materials, including strength and deformation characteristics, are summarized in Table 5.3. To determine key soil parameters such as the friction angle φ, cohesion c, Young's modulus E, and Poisson's ratio μ, triaxial compression tests were conducted under controlled conditions. In these tests, three experimental groups were selected, each subjected to a different confining pressure. For each group, five standard cylindrical soil samples (50 mm in diameter and 100 mm in height) were tested. The sample must have a uniform structure and be free of voids or defects to ensure reliable results. For cohesive soils or cases requiring saturated conditions, the sample is saturated before the test. This can involve placing the sample in a vacuum chamber to remove air or

Table 5.3 The mechanical parameters of slope in a saturated state

	Friction angle °	Cohesion kPa	Poisson's ratio	Young's modulus MPa	Permeability m/s	Density (Dry) kg/m^3
Colluvial soil	30	20	0.35	35	6×10^{-6}	1800
Weathered sandstone	32	85	0.30	55	1×10^{-6}	2100
Weak interlayer	20	15	0.36	32	6×10^{-6}	2000
Bedrock	36	100	0.13	300	5×10^{-7}	2200

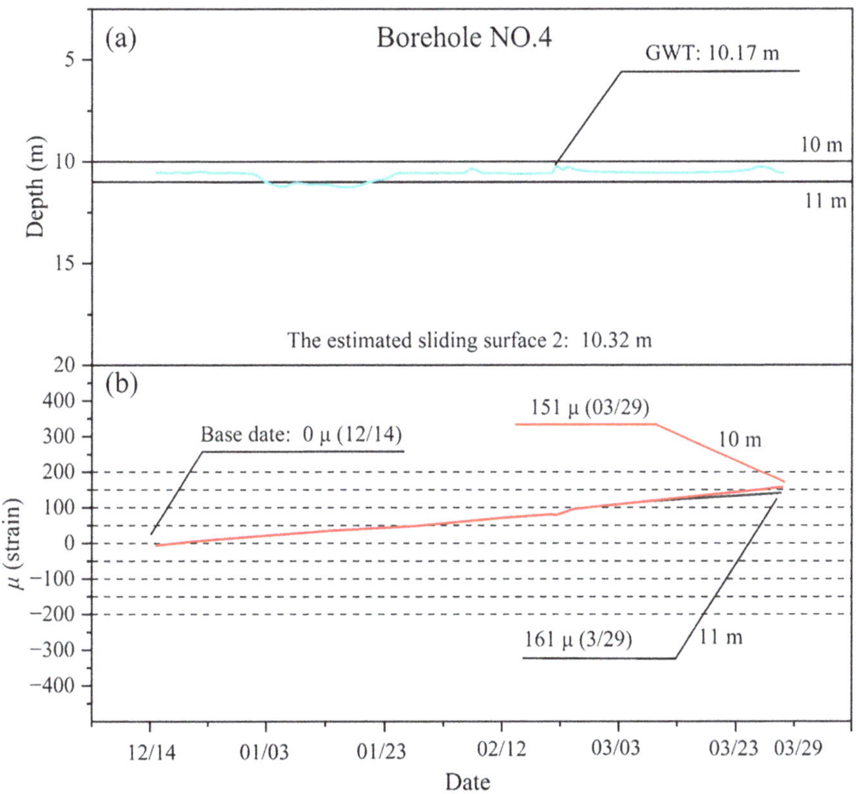

Fig. 5.7 Monitoring data of (**a**) groundwater table and (**b**) displacement in Borehole No. 4 from December 14, 2023 to March 29, 2024 (Li et al. 2025)

using backpressure saturation to achieve a high degree of saturation (typically over 95%). The degree of saturation is confirmed using Skempton's B-value test. The soil sample is encased in a thin, impermeable rubber membrane to prevent leakage of water or air. The membrane is secured using O-rings at both ends of the sample. The sample is placed on the pedestal of the triaxial cell, ensuring proper alignment. Soil samples are placed at the top and bottom of the sample to allow drainage and uniform load transfer. Drainage lines and pressure transducers are connected to the cell to facilitate pore pressure measurement and control. A controlled confining pressure is applied around the soil sample by filling the cell with water or air. This simulates in-situ lateral stresses. The confining pressure is gradually increased to the desired value while monitoring to ensure uniform application. The system is left to equilibrate, allowing the confining pressure to stabilize and any excess pore water pressures to dissipate (in drained conditions). Axial stress is applied to the sample via a piston or loading ram. The loading is conducted in undrained test (CU): Axial loading is faster, preventing pore pressure dissipation, replicating short-term conditions. Displacement transducers measure axial deformation, and volume changes (in drained conditions) are tracked via fluid displacement in the drainage system. Stress-strain data is continuously recorded. In CU tests, pore pressure transducers measure changes in pore water pressure as axial loading progresses. This provides insights into soil behavior under undrained conditions. The test concludes when the soil sample reaches failure, indicated by a peak in the stress-strain curve or the achievement of a specific strain level (e.g., 20% axial strain). Axial stress and strain data are recorded throughout the test, allowing determination of key soil parameters such as cohesion c, friction angle φ, and Young's modulus E. After the test, the soil sample is examined for failure mode, such as shear plane development or bulging. Stress-strain and pore pressure data are analyzed to derive strength parameters, such as shear strength and effective stress conditions. Results from multiple tests under varying confining pressures provide a Mohr-Coulomb failure envelope, which is essential for geotechnical design. The triaxial compression test provides critical insights into soil strength and deformation behavior under controlled stress states. By varying test conditions (drained or undrained), it accommodates diverse field scenarios, aiding in the design of foundations, slopes, and earth structures.

After applying the confining pressure, the specimens were compressed in the vertical direction while the stress-strain curves during the compression process were recorded. The internal friction angle φ and cohesion c were calculated based on the Mohr-Coulomb criterion. The Young's modulus was determined from the slope of the stress-strain curve in the elastic phase, while the Poisson's ratio was calculated from the ratio of lateral strain to axial strain. The average values of the results from all the experimental groups are the calculated parameter values shown in Table 5.3. This procedure ensures a comprehensive evaluation of shear strength and deformation behavior, providing critical inputs for accurate slope stability analysis and modeling. The dry density of the soil is determined using the Core Cutter Method, and the permeability coefficient of the soil is measured using a variable head permeability test (e.g., JGS0311), which has been illustrated in Sect. 2.2. From September 13th to 17th, 2023, the average maximum hourly rainfall during these 5 days is 20 mm/h. Thus, a constant rainfall rate of 20 mm/h was used in the calculation. As shown in Fig. 5.8b, the calculated safety factor of the slope was greater than 1 after 5 days of continuous rainfall when the homogeneous geotechnical parameters of the slope materials were determined. An important factor is that the heterogeneity of the slope material parameters was not considered which may lead to the physical parameters of the sliding zone being smaller than their mean values. For this mixture of colluvial soil and strongly weathered sandstone, the distribution of c and elastic modulus E was suggested to follow a Weibull distribution in this study. When the shape modulus m is taken as 1, it refers to a specific assumption regarding the distribution of material properties, particularly the cohesion c and Young's modulus E in the Weibull distribution. The shape modulus m is a parameter that controls the shape of the probability density function in the Weibull distribution. Different values of m lead to different distribution characteristics of the parameters of slope materials. When $m = 1$, the Weibull distribution simplifies to an exponential distribution. This implies that the cohesion and Young's modulus follow an exponential decay, where the likelihood of extreme values (either very high or very low) is more pronounced, and the material properties are assumed to be less uniformly distributed. Therefore, $m = 1$ represents the case where the physical and mechanical parameters of the rock sample are the most heterogeneous, with a broad range of variation in properties. This typically leads to lower overall strength and more complex deformation behavior in the rock sample. In the slope stability analysis, using $m = 1$ allows for a straightforward probabilistic model where the material properties (such as cohesion and Young's modulus) vary according to this simplified distribution. As a result, the recalculated slope safety factor may be lower than 1, indicating that the slope is less stable under the assumed distribution of material properties, as shown in Fig. 5.8a. Taking $m = 1$ provides a conservative estimate of the slope's stability by considering a less uniform distribution of the critical parameters. In

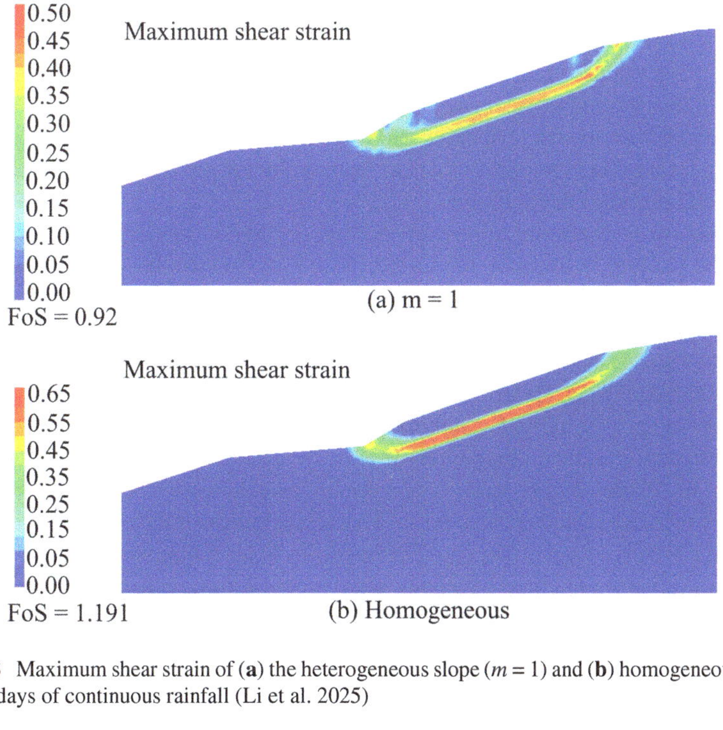

Fig. 5.8 Maximum shear strain of (**a**) the heterogeneous slope ($m = 1$) and (**b**) homogeneous slope after 5 days of continuous rainfall (Li et al. 2025)

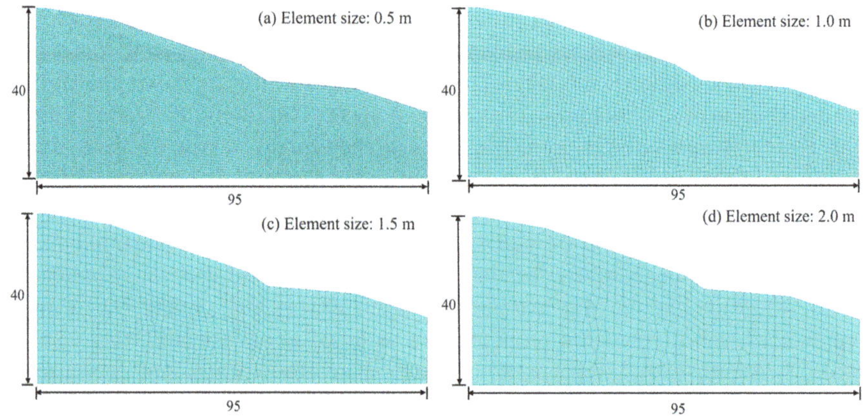

Fig. 5.9 The slope model is divided into grids of (**a**) 0.5 m and (**b**) 1.0 m, (**c**) 1.5 m, and (**d**) 2.0 m sizes, respectively (unit: m)

Table 5.4 FoS of the slope after 5 days of continuous rainfall with different element sizes

Element sizes (m)		2.0	1.5	1.0	0.5
Zones		715	1264	2625	10,637
Gridpoints		1560	2698	5502	21,782
FoS	$m = 1$	1.277	1.27	0.920	0.920
	Homogeneous	1.436	1.39	1.20	1.191

Fig. 5.8a, setting $m = 1$ in the Weibull distribution implies a specific assumption about the distribution of material properties. This distribution represents the most heterogenous mechanical properties and it influences the recalculation of the safety factor by introducing variability that can potentially reduce the margin of safety for the slope under rainfall conditions.

The results of the finite difference method (FDM) using the strength reduction technique are highly sensitive to mesh design. Previous studies indicate that hexahedral elements are preferable for meshes in models with simple geometries (Zhang et al. 2013). Accordingly, hexahedral elements were employed for mesh generation in this study. Mesh sensitivity analyses were conducted using different element sizes to evaluate their effects on the slope's factor of safety (FoS) under rainfall conditions (Fig. 5.9). Table 5.4 summarizes the number of grid points and zones for each mesh configuration, along with the corresponding FoS after 5 days of continuous rainfall. While smaller mesh sizes improve calculation accuracy, they also substantially increase computational costs. Therefore, optimizing mesh size is essential to balance precision and efficiency. The results in Table 5.4 indicate that the FoS for both homogeneous and heterogeneous slope models converges when element sizes are reduced to 1 m and 0.5 m. Based on this observation, a 1 m element size was selected for the Sasebo landslide slope model, providing a suitable compromise between accuracy and computational efficiency.

This chapter demonstrates that incorporating the heterogeneity of the weathered layer in slope stability analyses under rainfall conditions produces more accurate results. It better reflects observed slope behavior in the field. The findings show that the Weibull distribution provides a convenient and reasonably precise approach for characterizing the heterogeneity of weathered layers. Accordingly, selecting an appropriate heterogeneity index m is critical for reliable slope stability assessment. The study further highlights that the dissipation of matric suction within weathered layers is a primary factor driving the reduction of slope stability during intense rainfall. As rainfall continues, the unsaturated zone within the slope gradually diminishes, and matric suction dissipates. This process leads to a rapid decline in cohesion and elastic modulus in the weathered layers, which causes a marked reduction in slope stability during short-term heavy rainfall events. Overall, this research advances the understanding of unsaturated water infiltration in slopes and its impact on slope stability under extreme rainfall conditions.

5.2 Sasebo Landslide (3D)

As shown in Fig. 5.1a, the Digital Elevation Model (DEM) of the entire area was first generated. A square region with a side length of 150 m was then selected to encompass the landslide area, and the contours of this region were extracted in the ArcGIS software (Fig. 5.1b). These contours were imported into Rhino software to create a 3D terrain model. The resulting data were transferred into FLAC^{3D} to construct the 3D slope model, following the grid division of the 3D model (Fig. 5.10a). The model has a length of 150 m, a width of 150 m and a height of 40 m. It was divided into 495,000 zones and 524,423 gridpoints. Additionally, the groundwater level before rainfall was included to generate the initial seepage field, with a maximum matric suction of 50 kPa, as shown in Fig. 5.10b. A constant rainfall rate of 20 mm/h for 5 days was performed in the calculation. For this mixture of colluvial soil and strongly weathered sandstone, the distribution of c and elastic modulus E was suggested to follow a Weibull distribution in this study. In this numerical simulation, the physical and mechanical parameters of the weathered rock in the slope were dynamically adjusted in real-time according to changes in saturation during the rainfall infiltration process. Due to the differences in porosity between the weathered layers, the reduction factors λ for the physical and mechanical parameters of the strongly and moderately weathered layers in a saturated state were set to 0.8 and 0.7, respectively.

Figure 5.11 shows the initial saturation of the slope and the saturation after 48 h of rainfall. Figure 5.12 presents the maximum shear strain contour of the slope after 5 days of continuous rainfall. Compared to Fig. 5.1c, the actual failure area of the slope closely resembles the potential sliding area indicated by the numerical simulation results, suggesting that the selected calculation parameters are reasonable. Additionally, considering heterogeneity further enhances the accuracy of the

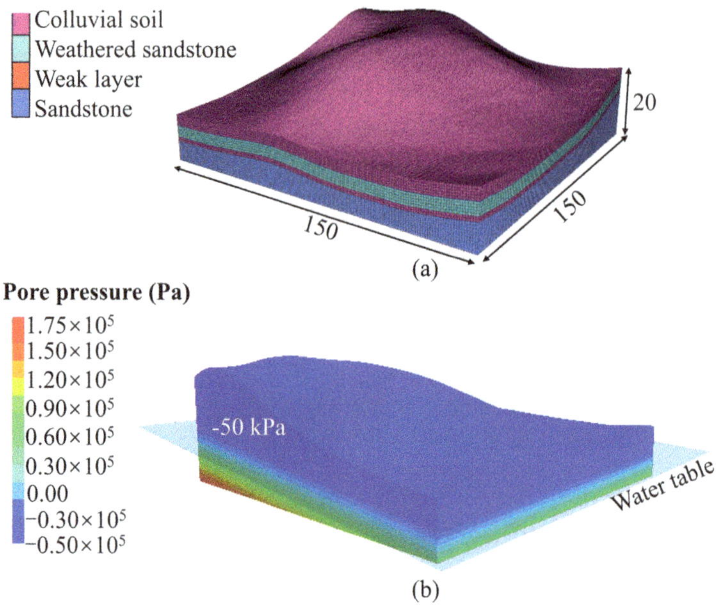

Fig. 5.10 (**a**) The 3D slope model of Sasebo landslide in FLAC^{3D} (unit: m). (**b**) The distribution of pore pressure before rainfall in the slope

Fig. 5.11 Distributions of saturation of the slope (**a**) before and (**b**) after 48 h of rainfall

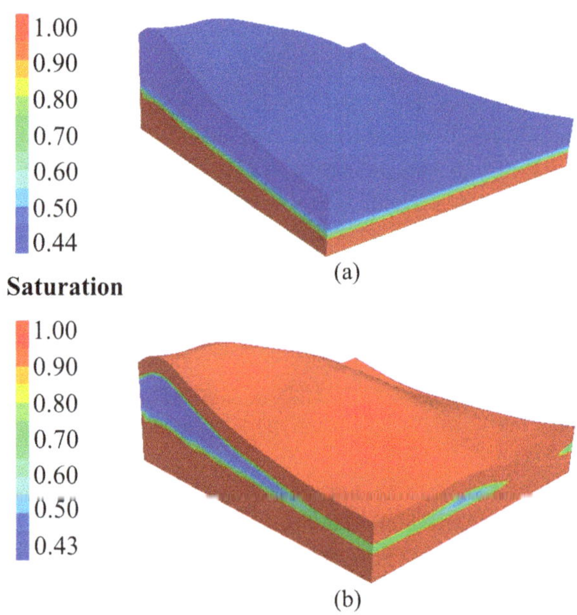

Fig. 5.12 Maximum shear strain contour map of the slope after 5 days of continuous rainfall

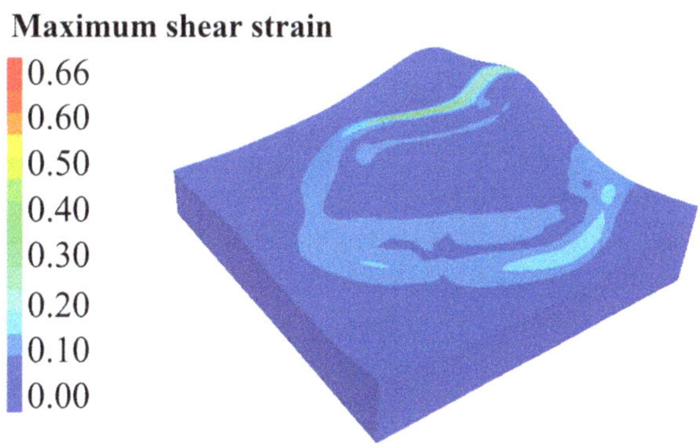

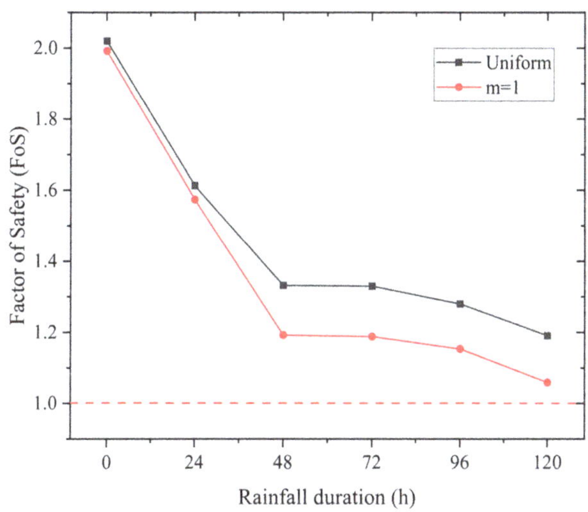

Fig. 5.13 Comparison of FoS for homogeneous and heterogeneous slopes under rainfall conditions

calculations. As shown in Fig. 5.13, the three-dimensional slope safety factor after 5 days of continuous rainfall is 1.06 by considering the heterogeneity of weathered layers, while the result from the homogeneous calculation is 1.19. It is clear that the safety factor obtained from the heterogeneous slope can reflect the actual slope failure situation. This demonstrates that dynamically adjusting the physical and mechanical parameters of the slope elements based on changes in saturation during rainfall is a more accurate and reasonable calculation method. Unlike the study by Pan et al. (2020), this research does not account for the effects of changes in the groundwater table on the slope stability. This is because the focus of this study is on the stability of shallow layers of rock slopes under heavy rainfall. However, due to the limitations of $FLAC^{3D}$ in computing preferential flow along interfaces, the impact of tension crack infiltration is not considered in this study.

5.3 Conclusions

Using the Sasebo landslide as a case study, this chapter highlights the importance of accounting for the heterogeneity of weathered layers in slope stability analysis under rainfall conditions, demonstrating that this approach yields more accurate results aligned with real-world scenarios. The findings emphasize that the Weibull distribution serves as a convenient and precise method for characterizing the heterogeneity of weathered layers. Consequently, selecting an appropriate heterogeneity index m is critical for accurately assessing slope stability. Due to the natural heterogeneity of rocks and the additional complexity introduced by joints and fractures, the weakening effect in rock masses can be quite intricate. This chapter employs FISH scripting within FLAC^{3D} for secondary development, using the Weibull distribution to model the heterogeneity of rocks. By incorporating experimental results from Pan et al. (2020), the FISH functions were utilized to simulate the water-softening effect in soft rocks. The approach involves adjusting the physical and mechanical parameters of the rocks in saturated conditions based on different porosities, thereby replicating the variations observed in uniaxial compression test results due to varying water content. In this study, numerical simulations of uniaxial compression tests are conducted to investigate the impact of rock heterogeneity on the peak compressive strength σ_{peak} of sandstone samples by considering various rock porosities and saturation levels. The research by Pan et al. (2020) is extended to heterogeneous rock slopes under heavy rainfall condition. The cohesion c and elastic modulus E of weathered rock masses are assigned using a Weibull distribution, with different degrees of rock heterogeneity defined by the homogeneity index m. This study incorporates the effects of matric suction in heavily and moderately weathered layers, as well as fluid-solid coupling in the 2D and 3D slope stability analysis. Different porosities were assigned to different weathered layers, and FISH functions were utilized to adjust the strength of elements based on varying degrees of saturation. In the 2D slope model, the porosity of the heavily weathered layer was set at 0.16, while the moderately weathered layer was set at 0.14. Since actual slopes experience gradual increases in saturation over 48 h of rainfall with initially low water content, conservative reduction factors of 0.8 and 0.7 were applied to the weathered layers, respectively. The accurate determination of reduction factors λ is closely related to the initial state of the slope and the duration of rainfall, necessitating further studies to refine these factors. Compared to studies that typically use a simplified approach distinguishing only between saturated and dry states during rainfall infiltration, it is evident that real-time adjustment of the physical and mechanical parameters of rocks based on changes in saturation during rainfall provides a more accurate and rational calculation method. Otherwise, it may lead to a significant overestimation of the slope's safety factor. Applying the same reduction factors λ to the actual 3D slope in Sasebo yielded a factor of safety (FoS) of 1.06, indicating a near-failure state that closely matches the observed slope failure condition. In contrast, the calculation for the homogeneous slope produced a factor of safety (FoS) of 1.191, indicating a relatively stable slope. This demonstrates the validity of the average parameter values obtained from experiments and highlights the significant impact of accounting for the heterogeneity of weathered layers in accurately assessing slope stability under rainfall conditions. Overall, this research significantly advances the understanding of unsaturated rainwater seepage processes in slopes during heavy rainfall events and their impact on slope stability.

References

Cai F, Ugai K (2004) Numerical analysis of rainfall effects on slope stability. Int J Geomech 4:69–78. https://doi.org/10.1061/(ASCE)1532-3641(2004)4:2(69)

Li X, Jiang Y, Sugimoto S (2025) Slope stability analysis under heavy rainfall considering the heterogeneity of weathered layers. Landslides 22:1257–1273. https://doi.org/10.1007/s10346-024-02404-8

Ng CWW, Shi Q (1998) A numerical investigation of the stability of unsaturated soil slopes subjected to transient seepage. Comput Geotech 22:1–28. https://doi.org/10.1016/S0266-352X(97)00036-0

Pan Y, Wu G, Zhao Z, He L (2020) Analysis of rock slope stability under rainfall conditions considering the water-induced weakening of rock. Computers and Geotechnics 128:103806. https://doi.org/10.1016/j.compgeo.2020.103806

Zhang Y, Chen G, Zheng L et al (2013) Effects of geometries on three-dimensional slope stability. Can Geotech J 50:233–249. https://doi.org/10.1139/cgj-2012-0279

Chapter 6
Two-Dimensional (2D) Versus Three-Dimensional (3D) Slope Stability

Abstract This chapter investigates the impact of complex geometries on three-dimensional (3D) slope stability under short-term heavy rainfall based on numerical simulations. Different groups of standard slope models were designed to examine the effects of surface undulation and slope gradient on stability of 3D slopes. For a 4 m surface undulation increment, the 3D slope exhibits lower safety factors than the selected two-dimensional (2D) critical section, indicating that 2D analyses are not necessarily more conservative when surface undulation is sufficiently large. Further analyses show that higher undulation frequency or degree further reduces 3D safety factors, with differences from 2D results becoming pronounced when the undulation exceeds 10 m. These findings highlight the limitations of traditional 2D critical section methods in capturing the effects of complex 3D geometries. Overall, the study demonstrates that 3D slope geometry significantly influences stability calculations under rainfall conditions. Performing actual 3D analyses is therefore essential for slopes with complex geometry. It is strongly recommended that engineers integrate both 2D and 3D assessments to ensure reliable and efficient slope design under short-term heavy rainfall.

Keywords Geometry characteristics · Surface undulation · Slope gradient · Standard 3D models · Heavy rainfall

6.1 Introduction

Two-dimensional (2D) slope stability analysis is commonly employed in numerical simulations due to its relative simplicity and fast computation. Under the plane strain assumption, three-dimensional (3D) slope stability can be approximated by a 2D analysis when one dimension of the slope is significantly larger than the other

The content of this chapter is based on a revised version of Li et al. (2024).

Y. Jiang, X. Li, *Stability Evaluation of Slope Under Heavy Rainfall Based on Geometric Features*, Springer Natural Hazards,
https://doi.org/10.1007/978-981-95-8656-1_6

two and the geometry lacks curvature or corners (Zhang et al. 2013). This simplification generally produces conservative results, as 2D analyses often yield lower factors of safety (FoS) than 3D analyses. However, field observations indicate that most real slopes and landslides exhibit inherently 3D behavior, with slip surfaces that are spherical, ellipsoidal, or otherwise complex. This makes the plane strain assumption rarely valid in real 3D slopes (Jaboyedoff et al. 2020). In addition, the assumption of uniform geometry along the direction perpendicular to the 2D plane is inconsistent with the complex 3D slope shapes found in nature (Azizian and Popescu 2006; Leong and Rahardjo 2012). Wines (2016) highlighted that 3D analyses can account for boundary conditions, material heterogeneity, and the orientation of geological structures along all three spatial dimensions (x, y, z). It provides a closer representation of real slope conditions. Advances in computational methods and hardware have made 3D slope stability analyses increasingly feasible and widespread (Griffiths and Marquez 2007; Wei et al. 2009; Zhang et al. 2015; Zhou et al. 2020). For instance, Zhang et al. (2013) analyzed the effects of curved surfaces and turning corners under three different boundary conditions. Their results demonstrated that curvature and convex arcs can significantly reduce the stability of steep slopes. Furthermore, the FoS obtained from 3D models often differs substantially from plane strain solutions. In some cases, considering the orientation of geological structures in 3D can result in a lower FoS than 2D analyses. This is particularly true when the principal structures are misaligned by more than 20–30° relative to the slope strike (Wines 2016).

In summary, 2D analyses are computationally efficient. However, they are not universally conservative, particularly for slopes with complex geometry perpendicular to the plane of interest. Studies indicate that 3D analyses are essential when soil and groundwater characteristics, boundary conditions, and geometric complexities (e.g., ridges or corners) interact. For example, Leong and Rahardjo (2012) reanalyzed a residual soil slope in Singapore and demonstrated that differences between 2D and 3D FoS are more pronounced for slopes with low groundwater tables. Another key factor contributing to discrepancies between 2D and 3D analyses is matric suction (negative pore pressure). Zhang et al. (2015) showed that including matric suction in unsaturated zones above the water table can produce differences in FoS between 2D and 3D analyses of up to 60%, with the largest discrepancies occurring in concave slopes. Rainfall infiltration further modifies matric suction and interacts with slope geometry, affecting water movement and stability. Topography itself influences infiltration and runoff by altering initial soil moisture, effective rainfall, and subsurface water movement (Talebi et al. 2008; Sabzevari and Noroozpour 2014; Wang and Chen 2021). For example, Wang and Chen (2021) have demonstrated that longitudinal curvature and plan shape of slopes strongly affect rainfall infiltration patterns. Xu et al. (2021) studied a simple 3D unsaturated vertical cut slope under variable rainfall. They found that the width-to-height ratio and rainfall type significantly influence FoS and critical failure patterns. Building on these insights, this chapter focuses on how rainfall infiltration interacts with 3D slope geometry to influence the development of critical failure surfaces and overall slope stability.

The most widely used approach for 3D slope stability remains the limit equilibrium method (LEM), which typically extends 2D slice methods into 3D column approaches (Zheng 2012; Tozato et al. 2022). However, 3D LEM relies on several assumptions and is often difficult to adapt to realistic boundary conditions in the third dimension (Zhang et al. 2022). A major limitation is the challenge of identifying the critical 3D failure surface, both in terms of its shape and location (Zhang et al. 2013). Another common approach is the limit analysis method (LAM), with various 3D models proposed in the literature (Michalowski 2010; Li et al. 2010). Based on the upper bound theorem, a 3D rotational failure mechanism can be formulated and an energy balance equation used to calculate the factor of safety (FoS) (Xu et al. 2021). Nevertheless, constructing 3D failure mechanisms for complex slopes remains challenging in LAM. Currently, the strength reduction method (SRM) has emerged as an effective technique for 3D slope stability assessment due to several advantages (Ugai and Leshchinsky 1995; Griffiths and Marquez 2007; Nian et al. 2012). By progressively reducing material strength until failure, SRM provides a realistic evaluation of slope stability (Zienkiewicz et al. 1975). This enables detailed analyses of complex slope geometries, spatially varying soil properties, and critical slip surfaces (Sun et al. 2017). Wei et al. (2009) reported that FoS and failure modes obtained using SRM are generally consistent with those from LEM across multiple 3D case studies. While Nian et al. (2012) and Zhang et al. (2013) conducted comprehensive SRM studies on slopes with varying curvatures, gradients, and boundary conditions, they primarily considered single convex or concave surfaces without addressing multiple turning corners along the slope's longitudinal length.

In this chapter, an elastoplastic finite difference method (FDM) combined with a strength reduction technique was employed to determine the FoS of 3D slopes under heavy rainfall. The focus is on examining the effects of different slope surface geometries on the stability of large-scale slopes under short-term intense rainfall. A series of 3D slopes with continuous convex and concave surfaces and varying numbers of turning corners were constructed. Numerical simulations were used to investigate their influence on slope stability and critical failure surfaces. Furthermore, by comparing the FoS of representative 2D cross-sections with the 3D slope FoS, the effects of spatial dimensionality on slope stability under heavy rainfall were evaluated. This approach provides a deeper understanding of the mechanisms controlling actual 3D slope failure under rainfall conditions.

6.2 Characteristics of the Study Area

The target site is a rocky slope along National Route 251, extending approximately 4500 m along the coastline from Unzen City to Minamishimabara City in Nagasaki, Japan. To comprehensively evaluate the safety of the steep slopes along this route, a series of detailed geological investigations were conducted, including borehole drilling, standard penetration tests, in situ tests, and triaxial compression tests. The

Table 6.1 The material properties of the slope in the saturated state

	Friction Angle °	Cohesion kPa	Poisson's Ratio	Density kg/m³	Young's Modulus MPa	Permeability m/s
Strongly weathered sandstone	38	60	0.36	1800	200	6×10^{-6}
Moderately weathered sandstone	40	80	0.30	2100	300	1×10^{-6}
Bedrock	42	100	0.30	2200	600	5×10^{-7}

results of these surveys are summarized in Table 6.1. The weathered zone in the target area is approximately 6–8 m thick. The surface layer consists of colluvial soil mixed with strongly weathered sandstone, and its thickness was set to 3 m in this analysis. The second layer comprises moderately weathered sandstone and mudstone, and its thickness was also defined as 3 m. Beneath these layers lies the bedrock, which mainly consists of fresh sandstone. During short-term heavy rainfall, water infiltration causes significant softening of these rocks, leading to a reduction in their physical and mechanical parameters (Pan et al. 2020). To account for this effect, an empirical reduction factor of 0.8 was applied in the numerical simulations when the slope materials were considered fully saturated (Xu et al. 2017).

Given that the stability of these coastal slopes directly influences the design of protective measures for the coastal road, the application of 3D slope stability analysis is essential for ensuring roadway safety. Figure 6.1a shows the location of the road along the coastline and the digital elevation model (DEM) of the study area. The red box in Fig. 6.1a marks the study area of interest. Slope gradient is a key factor influencing slope stability (Rahardjo et al. 2007). To obtain a preliminary assessment of the stability of slopes along the coastal region, a representative area was selected, and slope gradients were extracted using grid-based mapping in GIS (Fig. 6.1b). The results indicate that the slope gradients along the coastline are predominantly around 50°, falling within a high-risk range that warrants further stability evaluation. The steepest sections (exceeding 50°) are concentrated along the coastal slopes, suggesting the need for more detailed analysis of their stability. Two specific areas with relatively steep slopes, referred to as Areas 1 and 2, were selected for detailed follow-up analysis (Fig. 6.1b). Figure 6.1c, d present the three-dimensional FLAC^{3D} numerical models for Areas 1 and 2, respectively. The models were constructed based on actual geological information, including groundwater level, slope height, and slope angle. Because the study focuses on shallow slope stability under rainfall conditions, variations in the groundwater table are not considered at this stage. Instead, only the surface undulations of the slopes were standardized to examine the influence of slope geometry on 3D stability.

Figure 6.1b shows two selected sections within a 420 m × 200 m area, both with slope heights of approximately 80 m. The key difference between the two sections lies in the degree of slope-surface undulation. Area 1 has a relatively smooth surface, whereas Area 2 exhibits much stronger undulation. In this study, the number of "turning corners" is used to represent different undulation frequencies of the

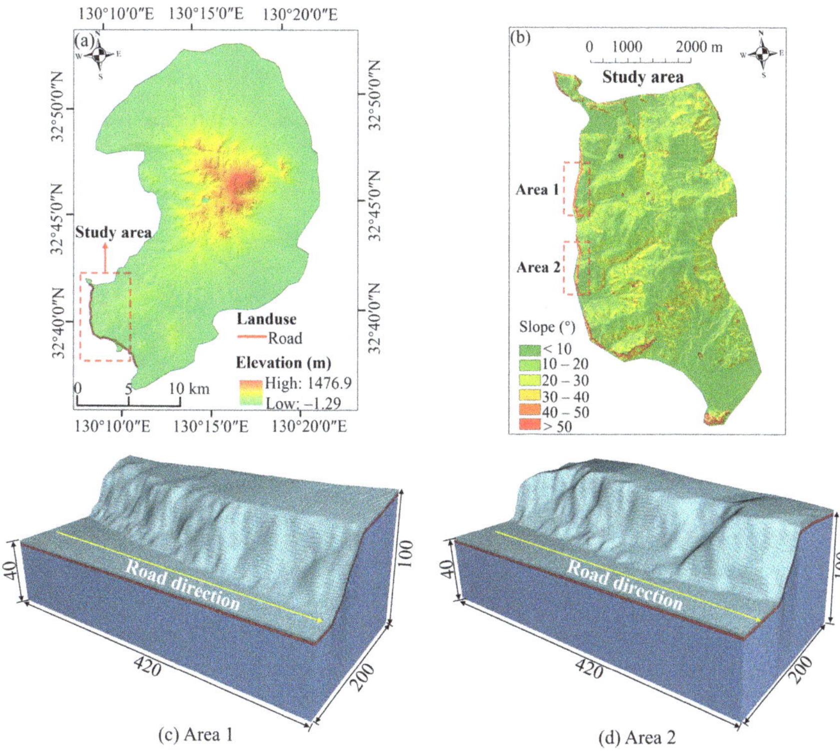

Fig. 6.1 (**a**) The position of the city road and the study area is within the red dashed lines. (**b**) The slope gradient statistics are derived from the grid mapping units, and the slope gradient is mainly around 50°

slope surface. To investigate how complex slope geometries affect 3D stability under heavy rainfall, a series of models were developed by assigning varying degrees and frequencies of surface undulation along a specified longitudinal length. Depending on the slope orientation relative to the road, the slopes were classified as either "parallel to the road" or "oblique to the road". For slopes oblique to the road, the longitudinal length of the intersecting cross-section increases progressively, as illustrated in Area 1 (Fig. 6.1c). In contrast, Area 2 shows a slope orientation that is nearly parallel to the road, combined with pronounced surface undulation (Fig. 6.1d).

6.2.1 Slopes Parallel to the Road

The first group of models is categorized as an orientation of slopes parallel to the road based on the surface geometry of Area 1 (Fig. 6.2). Four standard slope examples with different degrees of undulation were constructed in this section

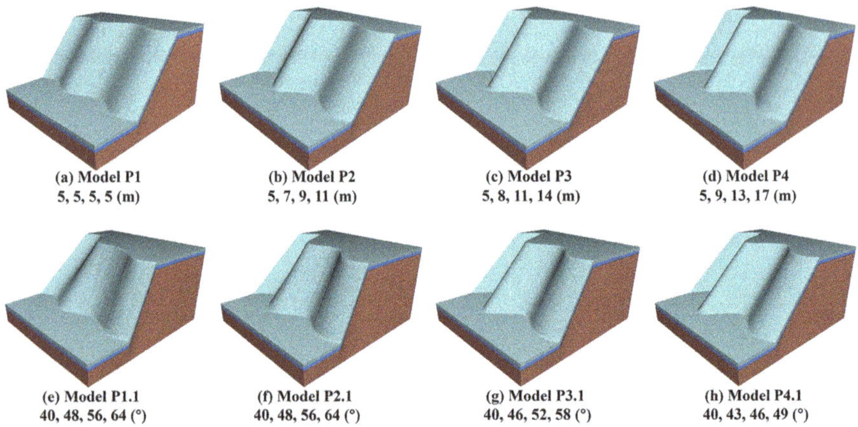

Fig. 6.2 The 3D models in Group 1, including Model (**a**) P1, (**b**) P2, (**c**) P3, and (**d**) P4, are characterized by a constant slope angle of 50°. Each model exhibits different configurations of slope surface undulations (unit: m). The 3D models in Group 2, including Model (**a**) P1.1, (**b**) P2.1, (**c**) P3.1, and (**d**) P4.1, are derived from the corresponding models in Group 1 but incorporate variations in slope angle

(Fig. 6.2a–d). Models P1 to P4 share identical overall slope dimensions, and differ only in the degree of surface undulation. The undulation increments for these four models are 0 m, 2 m, 3 m, and 4 m, respectively. Slope model P1 features a constant convex–concave amplitude of 5 m (Fig. 6.2a), while slope model P2 exhibits a linearly increasing undulation amplitude from 5 m to 11 m, with an interval of 2 m (Fig. 6.2b). Slope model P3 has an undulation amplitude that increases linearly from 5 m to 14 m at an interval of 3 m (Fig. 6.2c). Slope model P4 has an undulation amplitude that increases linearly from 5 m to 17 m at an interval of 4 m (Fig. 6.2d). All these four models are inclined at a slope angle of 50°, and homogeneous material properties are assigned to each weathered layer. This is because this chapter focuses on evaluating the influence of complex geometric configurations on 3D slope stability.

Figure 6.3a shows the top view of Model P2 with a continuous convex- and concave-shaped surface geometry, and the dashed line represents the baseline of the slope surface in the figure. The degree of undulation is defined as the distance between the undulation peaks and the baseline. The baseline is divided into four equal parts, with each segment serving as the longitudinal length of the undulation shape. For this slope model, the degree of undulation increases linearly from 5 m to 11 m with an even spacing of 2 m. Figure 6.3b shows the side view of the slope model, where AB represents the groundwater level and ABCD represents the unsaturated zone above the groundwater with a maximum matric suction of −50 kPa. As pointed out by Rahardjo et al. (2007), the length of the slope crest has a minimal impact on slope stability once the slope gradient and height are determined. Therefore, it is reasonable to select this 2D cross-section as the critical cross-section of 3D slope model.

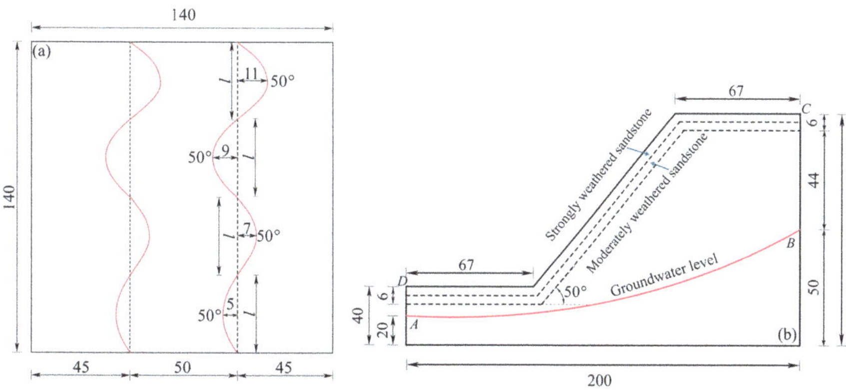

Fig. 6.3 (**a**) Top view of Model P2 (unit: m). (**b**) Side view of Model P2, and AB is the groundwater level, and ABCD is the unsaturated zone. (unit: m) (Li et al. 2024)

This chapter first examines how different degrees of surface undulation affect the stability of slopes that are parallel to the road, assuming a slope gradient of 50°. It then evaluates the influence of varying slope gradients using the same geometric configurations. In natural slopes, not only the surface curvature but also the slope gradient varies along the longitudinal direction. Therefore, this section investigates the influence of slope gradient variations on 3D slope stability while considering changes in slope-surface geometry. To achieve this, a second group of models, P1.1, P2.1, P3.1, and P4.1 (Fig. 6.2e–h), was developed based on the original Models P1, P2, P3, and P4. For Model P1.1, the slope gradient increases from 40° to 64° in increments of 8° for each convex and concave segment (Fig. 6.2e). Model P2.1 follows the same gradient pattern, with variations from 40° to 64° at 8° intervals (Fig. 6.2f). In Model P3.1, the gradient ranges from 40° to 58° with increments of 6° (Fig. 6.2g), whereas in Model P4.1, it varies from 40° to 49° with increments of 3° (Fig. 6.2h). This design allows comparison with the previous set of models, particularly in cases where slope-surface undulation is pronounced but variations in slope gradient are relatively small. Such comparison helps identify whether undulation amplitude or gradient variation has a greater influence on 3D slope stability. It is important to note that Model P4.1 maintains slope gradients consistently below 50°, whereas the other models include gradients both above and below 50°.

Figure 6.4a illustrates the top view of Model P2.1, in which the slope gradient increases from 40° to 64° in increments of 8° at each convex and concave segment. These angles correspond to the slope inclinations at the peak positions of the surface undulations, while the slope gradient remains 50° at the curvature inflection points. In the first convex segment, the gradient gradually increases from 50° to 64°, then decreases back to 50°. In the following concave segment, the gradient increases from 50° to 56°, then returns to 50°. By repeating this pattern, a series of 3D models with varying longitudinal gradient profiles was constructed. Figure 6.4b shows the side view of Model P2.1, representing its critical 2D cross-section. Figure 6.5 presents the side views of Models P3.1 and P4.1, which represent the critical 2D cross-sections of these two models.

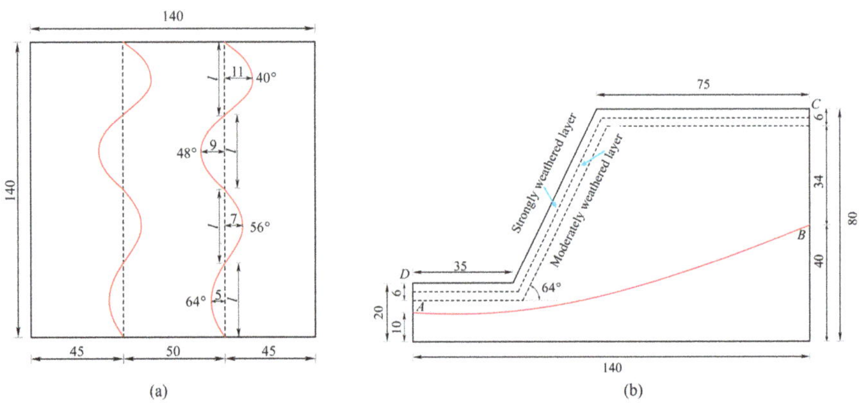

Fig. 6.4 (**a**) Top view of Model P2.1 (unit: m). (**b**) Side view of Model P2.1, and AB is the groundwater level, and ABCD is the unsaturated zone. (unit: m) (Li et al. 2024)

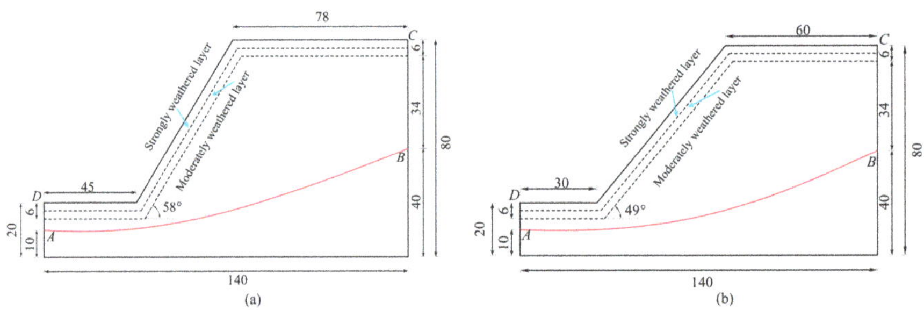

Fig. 6.5 (**a**) Side view of Model P3.1 with a slope gradient of 58° (unit: m). (**b**) Side view of Model P4.1 with a slope gradient of 49°, AB is the groundwater level (unit: m) (Li et al. 2024)

6.2.2 Slopes Oblique to the Road

Further analysis is conducted for slopes oriented obliquely to the road by examining how varying numbers of turning corners along the longitudinal direction affect 3D slope stability. Based on the previous analysis, the final two groups of models focus on slopes oblique to the road alignment. Five slope configurations with different undulation frequencies and amplitudes are evaluated, while the slope gradient is kept constant at 50° throughout this section. All FlAC^{3D} calculation models for the five 3D slope configurations are shown in Fig. 6.6. Figure 6.7a presents the top and side views of Model O3, with the dashed line indicating the baseline of the undulations. Models O1, O2, and O3 are categorized according to their undulation frequencies 2, 4, and 6, respectively. It represents the number of segments into which the baseline is divided (Fig. 6.7a). Since the slope gradient and height of the 3D models are fixed, a 2D cross-section is sufficient to represent the critical failure surface for all these models, as illustrated in Fig. 6.7b. Three slope models with

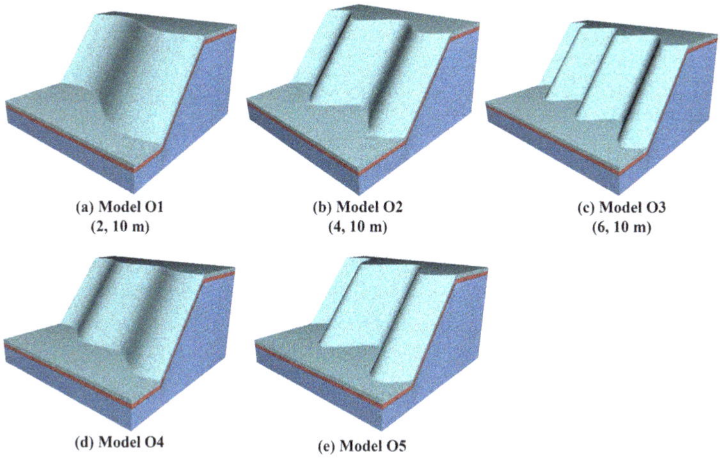

Fig. 6.6 (**a**) Model O1 with an undulation frequency of 2, (**b**) Model O2 with an undulation frequency of 4, (**c**) Model O3 with an undulation frequency of 6, (**d**) Model O4 with an undulation amplitude of 5 m, and (**e**) Model O5 with an undulation amplitude of 15 m

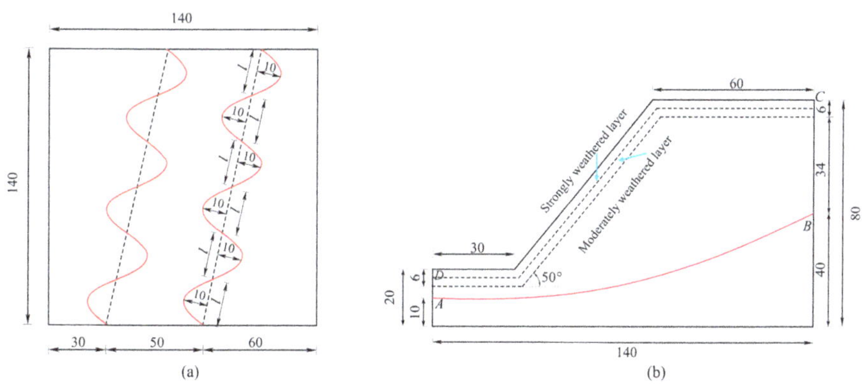

Fig. 6.7 (**a**) Top view of Model O3 (unit: m). (**b**) Side view of Model O3 with a gradient of 50°, and AB is the groundwater level, and ABCD is the unsaturated zone (unit: m) (Li et al. 2024)

different undulation frequencies constitute Group 3. The undulation amplitude, defined as the vertical distance between the undulation peak and the baseline, is used to construct two additional slope models: O4 and O5, with undulation degrees of 5 m and 15 m, respectively. Both models share the same undulation frequency of 4, consistent with Model O2. These models with different undulation degrees form Group 4.

6.3 Methodology

6.3.1 Unsaturated Seepage Analysis in FLAC^{3D}

In this study, a fluid–solid coupling framework based on the theory of continuous porous media was adopted to simulate unsaturated seepage within the slope. Unsaturated flow is achieved by enabling the hydraulic conductivity in the unsaturated zone to update automatically as pore pressure evolves. The detailed implementation of unsaturated seepage in FLAC^{3D} has been presented in Chaps. 3 and 4. Therefore, only the results of three-dimensional simulation are reported here. Regarding the initial matric suction distribution prior to rainfall, Leong and Rahardjo (2012) indicated that assuming a constant matric suction above the maximum suction value can improve the accuracy of rainfall infiltration simulations. Following this recommendation, and referencing Zhang et al. (2015), the maximum matric suction in this study was set to −50 kPa (Fig. 6.8). By applying a negative pore pressure of −50 kPa across the designated portion of the slope using FISH functions, the seepage equilibrium analysis produced a continuous negative pore-pressure field extending from the groundwater table to the slope surface (Fig. 6.8).

Rainfall data from Unzen County, Nagasaki City, Japan, are shown in Fig. 6.9. During the event from 11 to 13 August 2021, a total of 900 mm of precipitation was recorded, representing an exceptionally high cumulative rainfall. The maximum hourly intensity reached 81 mm/h, indicating extremely heavy rainfall over the three-day period. In the numerical simulation, a continuous rainfall duration of 48 h was applied, with a constant rainfall intensity of 20 mm/h, based on the saturated hydraulic conductivity of the slope. The dynamic rainfall boundary condition was implemented using FISH functions in FLAC^{3D}, while the pore pressure at the slope surface was set to zero (Fig. 6.10). Figure 6.10 illustrates the pore-pressure distribution in Model P2.1 after 24 and 48 h of rainfall. It can be clearly observed that as rainfall continues, infiltration depth increases and the negative pore pressure in the

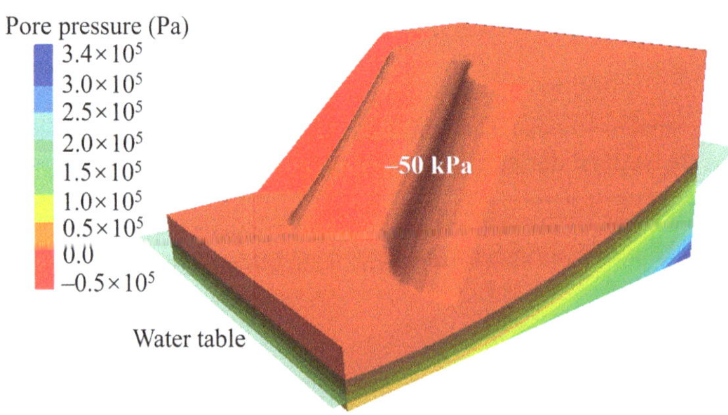

Fig. 6.8 The distribution of pore pressure (Pa) in the slope before rainfall

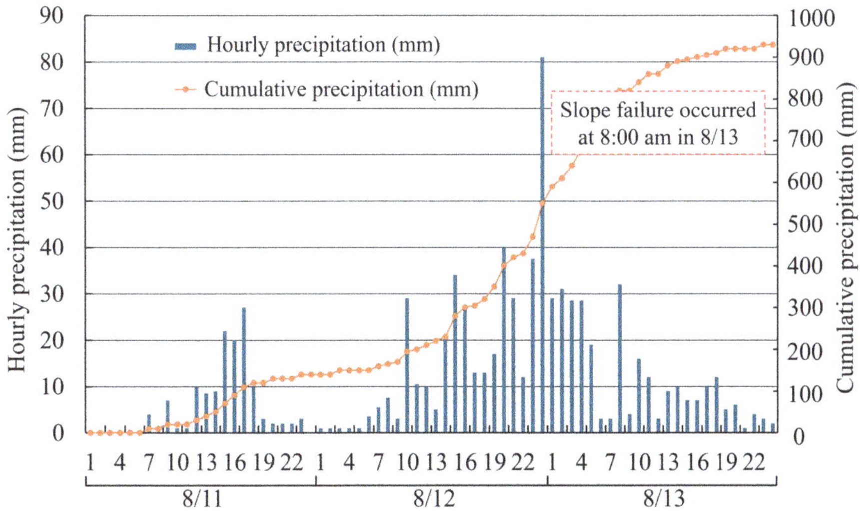

Fig. 6.9 Hourly and cumulative precipitation in Unzen county from 11 to 13 August 2021 (Li et al. 2024)

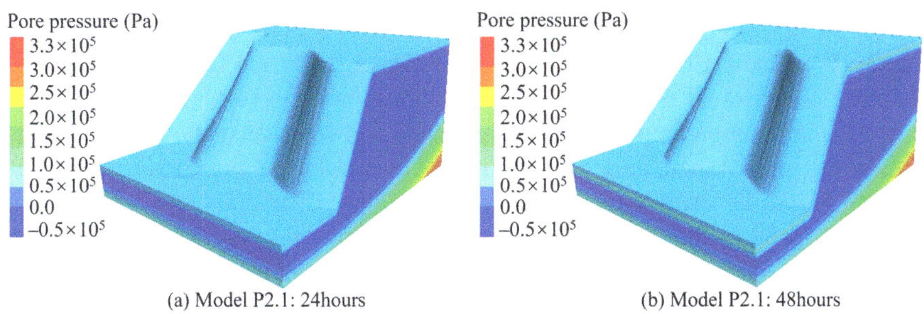

Fig. 6.10 The pore pressure distribution in the slope after 24 h (**a**) and 48 h (**b**) of rainfall. The pore pressure on the slope surface was set to 0 when surface runoff formed

unsaturated zone gradually transitions to positive pore pressure. The slope factor of safety was monitored at three critical time points: before rainfall, after 24 h of rainfall, and after 48 h of rainfall.

6.3.2 FDM with the Strength Reduction Method

The finite difference method (FDM) was chosen for its notable advantages in handling complex geological conditions and large-scale simulations. FDM efficiently accommodates heterogeneous materials and intricate boundary conditions, which is essential for accurately modeling water–rock interactions. Its computational efficiency makes it particularly suitable for large-scale three-dimensional simulations requiring high spatial and temporal resolution. Moreover, FDM exhibits strong numerical stability, mitigating divergence issues associated with complex boundaries and nonlinear behavior. The numerical simulations were performed using the FDM software Fast Lagrangian Analysis of Continua in three dimensions ($FLAC^{3D}$). The analyses employed an elastic, perfectly plastic material model governed by the Mohr–Coulomb yield criterion with an associated flow rule, where the tensile strength was not considered. In $FLAC^{3D}$, convergence is assessed using the nodal unbalanced force, defined as the sum of forces exerted on a node by its neighboring elements (Dawson et al. 1999). A node is considered in equilibrium when these forces sum to zero. In this study, the maximum unbalanced force across all nodes was normalized by the gravitational body force acting on each node. Convergence was achieved when the normalized unbalanced force for every node fell below 10^{-5}. The application of the shear strength reduction method is expressed in Eq. (3.55).

6.4 Results

6.4.1 Slopes Parallel to the Road

Figures 6.11 and 6.12 illustrate the maximum shear strain and factors of safety for the three-dimensional slopes in Group 1 before rainfall and after 48 h of rainfall. When the undulation amplitude of each convex and concave segment is 5 m, corresponding to zero increment in amplitude (0 m increment), the potential slip surfaces in Model P1 involve the entire slope both initially and after 48 h of rainfall (Figs. 6.11a and 6.12a). With an undulation increment of 2 m, the maximum shear strain begins to localize in the most prominent areas of the slope after 48 h of rainfall (Fig. 6.12b). As the undulation increment increases further to 3 m and 4 m, the initial potential slip surfaces are concentrated in the convex regions of the slope surface (Fig. 6.11c, d). However, as rainfall progresses, these slip surfaces gradually shift toward the concave regions of the slope (Fig. 6.12c, d). Overall, Fig. 6.12a, b show that after 48 h of rainfall, the potential slip surfaces are predominantly located between the most convex and concave peaks. This indicates that the most hazardous areas correspond to regions with the highest degrees of convexity and concavity. This spatial distribution differs from the initial hazardous zones observed before rainfall (Fig. 6.11a, b). Initially, slopes with undulation amplitude increments of 0 m and 2 m tend to experience overall failure, whereas slopes with increments of 2 m, 3 m, and 4 m exhibit potential slip surfaces concentrated in the most prominent areas. After 48 h of rainfall, these surfaces remain primarily located in the convex regions of the slope.

Figures 6.13 and 6.14 show the maximum shear strain and factors of safety for the 3D slopes in Group 2 before rainfall and after 48 h of rainfall. When variations in slope gradient are incorporated into the slope geometry, the areas of maximum

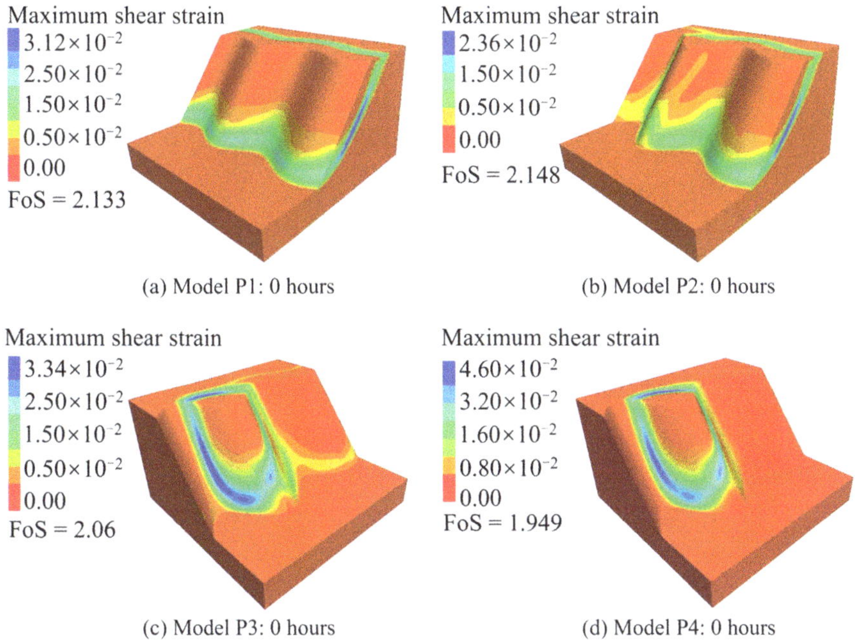

Fig. 6.11 The maximum shear strain contours and factors of safety for Models (**a**) P1, (**b**) P2, (**c**) P3 and (**d**) P4 before rainfall

shear strain concentrate in the steepest regions of the slope whenever the gradient exceeds 50° (Fig. 6.13). In contrast, in Model P4.1, where the undulation amplitude is maximal but the slope gradient remains below 50°, the potential sliding surfaces are located in the most prominent areas rather than in the steepest section (49°), as shown in Figs. 6.13d and 6.14d. This observation suggests that variations in slope gradient exert a stronger influence on slope stability than changes in surface undulation. In three-dimensional slope models, sliding surfaces are most likely to develop in areas of sharp curvature change, typically between the most convex and concave regions. This is because external forces such as rainfall infiltration, groundwater flow, or seismic activity induce stress concentration in these zones, increasing the likelihood of failure. Convex areas, with their higher curvature, amplify stress concentration and are more prone to shear failure. Concave regions similarly accumulate stress, particularly under rainfall-induced pore-pressure changes, which further destabilizes the slope. The transition zones between convex and concave areas experience the greatest stress variation, making them especially susceptible to sliding. Understanding the interaction between surface curvature, stress distribution, and external forces is therefore essential for accurately predicting slope failure and improving landslide risk assessment and mitigation strategies.

Table 6.2 summarizes the stability of three-dimensional slopes with varying undulation amplitudes and slope gradients for Groups 1 and 2. When the undulation

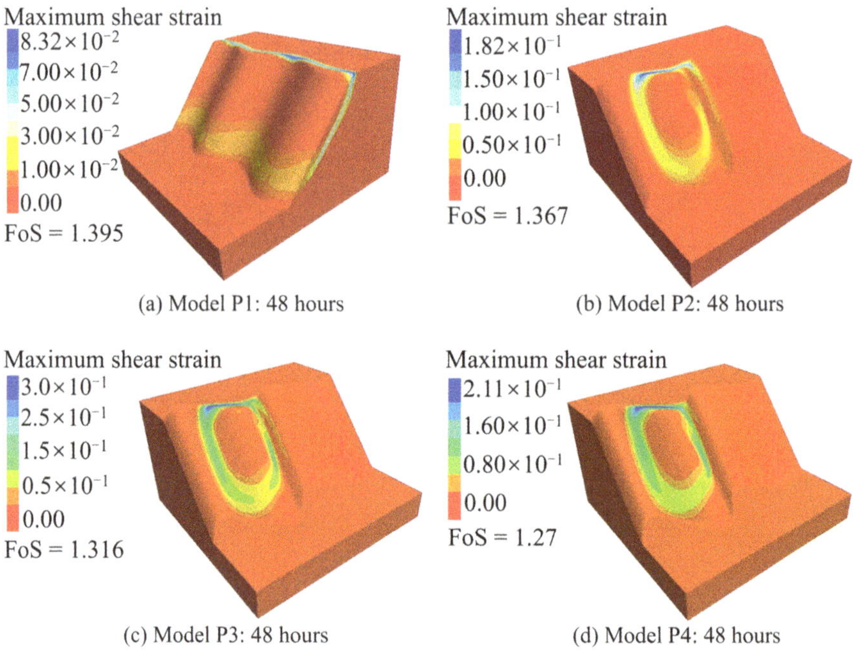

Fig. 6.12 The maximum shear strain contours and factors of safety for Models (**a**) P1, (**b**) P2, (**c**) P3 and (**d**) P4 after 48 h of rainfall

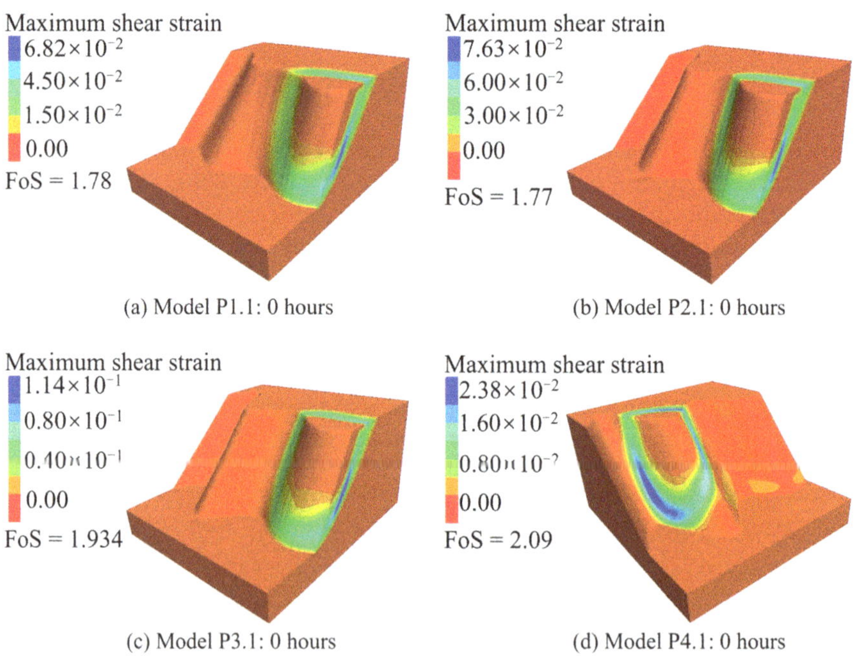

Fig. 6.13 The maximum shear strain contours and factors of safety for Models (**a**) P1.1, (**b**) P2.1, (**c**) P3.1 and (**d**) P4.1 before rainfall

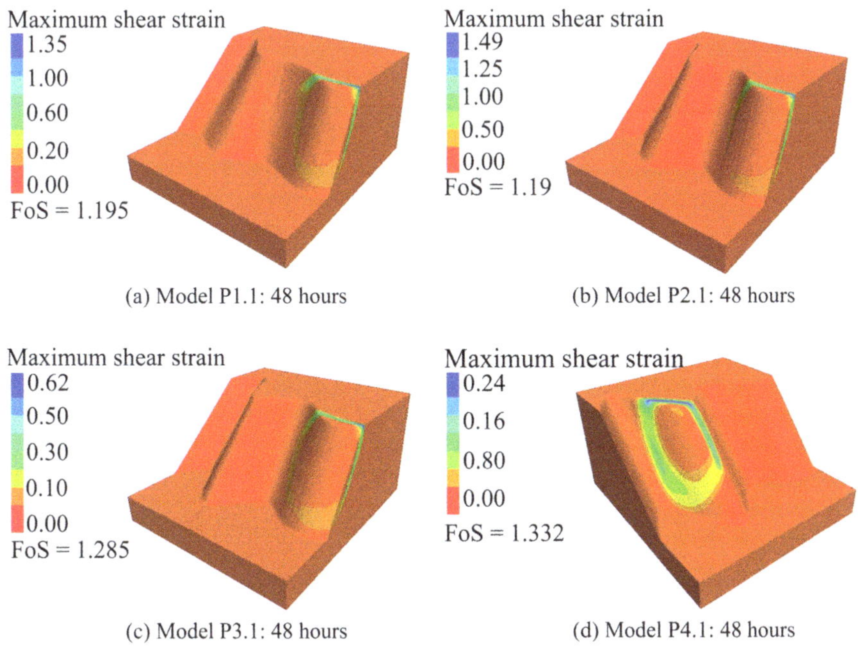

Fig. 6.14 The maximum shear strain contours and factors of safety for Models (**a**) P1.1, (**b**) P2.1, (**c**) P3.1 and (**d**) P4.1 after 48 h of rainfall

Table 6.2 Stability of 3D slopes with different undulation degrees and slope gradients

Group	Model	Undulation degrees (m) Slope gradient (50°)	FoS Rainfall duration (h) 0 h	24 h	48 h	DF (%) 0 h	24 h	48 h
1	P1	5, 5, 5, 5	2.133	1.629	1.395	3.54	0.99	9.84
	P2	5, 7, 9, 11	2.148	1.613	1.367	4.27	0.00	7.64
	P3	5, 8, 11, 14	2.06	1.50	1.316	0.00	−7.01	3.62
	P4	5, 9, 13, 17	1.949	1.42	1.27	−5.39	−11.97	0.00
2D analysis (slope gradient: 50°)			2.06	1.613	1.27			
		Variation of slope gradients (°)						
2	P1.1	40, 48, 56, 64	1.78	1.340	1.195	12.23	5.51	16.02
	P2.1	40, 48, 56, 64	1.77	1.33	1.19	11.60	4.72	15.53
	P3.1	40, 46, 52, 58	1.934	1.461	1.285	9.27	3.03	13.42
	P4.1	40, 43, 46, 49	2.09	1.527	1.332	2.50	−5.16	5.55
2D analysis		49	2.039	1.61	1.262			
		58	1.77	1.418	1.133			
		64	1.586	1.270	1.03			

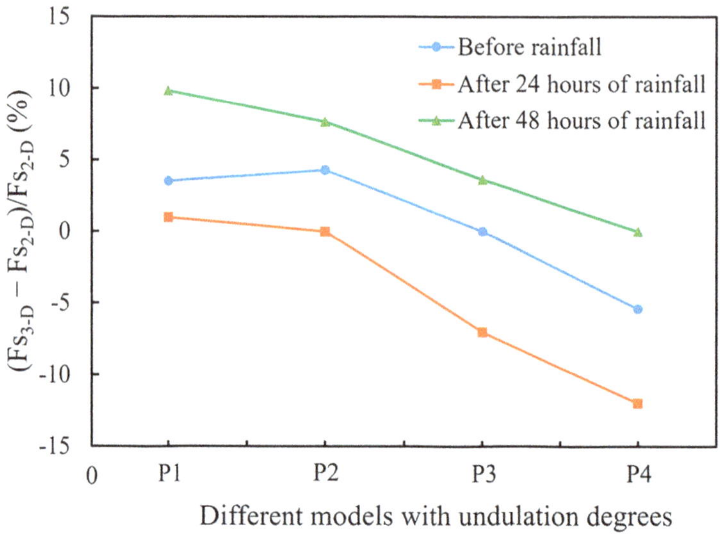

Fig. 6.15 Difference in the FoS for different increments of undulation degrees (Li et al. 2024)

frequency is four and the slope gradient is 50°, larger undulation amplitude increments correspond to lower slope factors of safety (FoS) after 24 and 48 h of rainfall. For 2D cross-sections, the FoS is lower than that of 3D slopes only when the undulation amplitude increments are 0 m or 2 m. This indicates that when the slope undulation is significant (e.g., 4 m increment), the safety factor of the 2D critical cross-section does not accurately represent that of the 3D slope. Therefore, 2D calculations are not necessarily more conservative than 3D analyses when the slope geometry is complex, highlighting the necessity of 3D FoS evaluation in such cases. When slope gradient variations are included, the FoS of the 2D cross-section with the steepest gradient is lower than that of the corresponding 3D slope. This makes the 2D calculation more conservative. However, identifying the 2D cross-section with the steepest gradient along a longitudinal slope in practice is challenging. Selecting a cross-section with a 50° slope gradient may underestimate the stability of 3D slopes, potentially leading to misjudgment of slope stability. Conversely, choosing the steepest 2D cross-section (64°) may result in overly conservative design. Table 6.2 presents the differences in factors of safety (DF) between two-dimensional (2D) and three-dimensional (3D) slope stability analyses ($(Fs_{3\text{-}D} - Fs_{2\text{-}D})/Fs_{2\text{-}D}$, or $\Delta Fs/Fs_{2\text{-}D}$).

Figure 6.15 shows that as the undulation degree of the slope surface increases, the factor of safety (FoS) of the three-dimensional slope decreases under the same rainfall duration. The difference in safety factors (ΔFoS) between 2D and 3D analyses remains positive for undulation increments of 0 m, 2 m, and 3 m. It indicates that 2D calculations are more conservative in these cases. However, when the undulation increment reaches 4 m, the differences in FoS before rainfall and after 24 h of rainfall are −5.39% and −11.97%, respectively. This demonstrates that at high

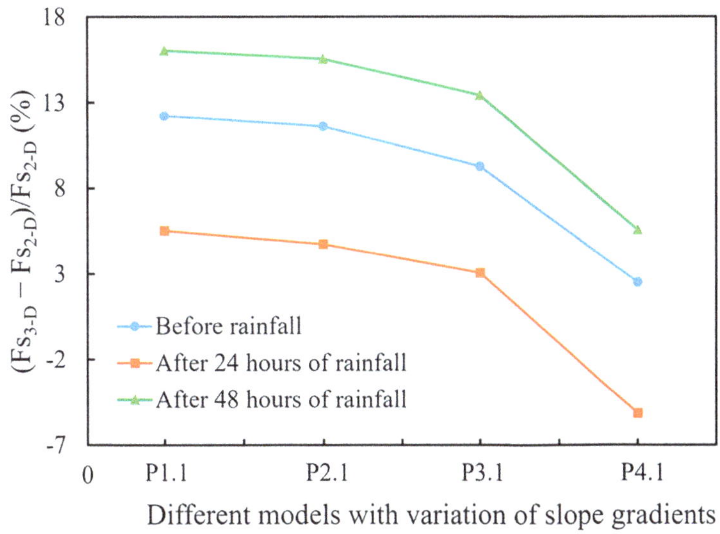

Fig. 6.16 Difference in the factor of safety for different increments of slope gradients (Li et al. 2024)

undulation levels, 3D results may be lower than 2D results, meaning the 2D analysis is no longer conservative. Similarly, Fig. 6.16 shows that with varying slope gradients, the FoS of 3D slopes gradually decreases as undulation increases after 24 and 48 h of rainfall. Since the most critical 2D cross-section (with the maximum slope gradient) was considered, the DF between 2D and 3D analyses remains mostly positive. This confirms that 2D calculations are generally more conservative. This is consistent with traditional 2D plane strain assumptions, which consider the slope as having an infinite longitudinal extent. For example, the entire slope is assumed to have a gradient of 64° in a 2D model, which is less stable than a 3D slope with gradients varying from 40° to 64°. Specifically, for Model P1.1, the DF between 2D and 3D slopes before rainfall and after 48 h of rainfall are 12.23% and 16.02%, respectively. For Model P2.1, the corresponding differences are 11.60% and 15.53%, respectively. Although selecting the 2D cross-section with the maximum gradient provides conservative safety estimates, it may lead to overly cautious and inefficient slope designs. Therefore, conducting a 3D slope stability analysis allows for a more rational and efficient approach to slope safety assessment.

6.4.2 *Slopes Parallel to the Road*

Figures 6.17a–c illustrates the maximum shear strain and factors of safety (FoS) for the three-dimensional slopes in Group 3 after 48 h of rainfall. When the undulation amplitude is 10 m, the post-rain FoS decreases as the undulation frequency increases.

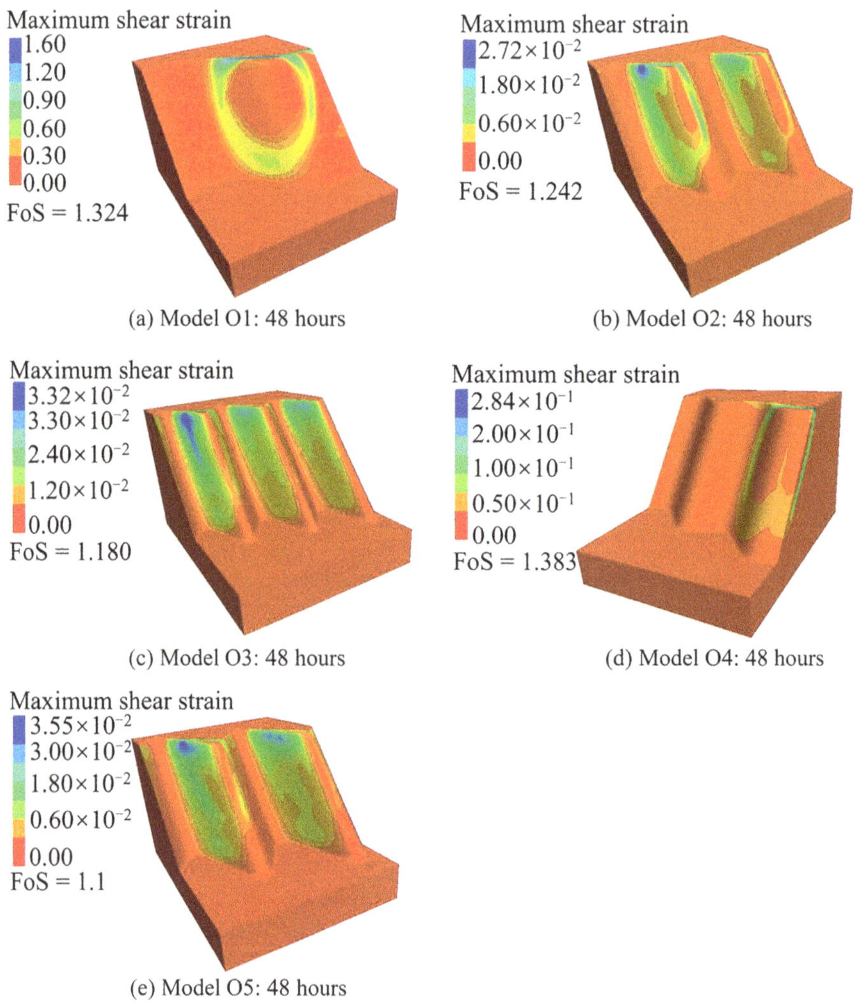

Fig. 6.17 The maximum shear strain contours and factors of safety for Models (**a**) O1, (**b**) O2, (**c**) O3, (**d**) O4 and (**e**) O5 after 48 h of rainfall

In general, the FoS of the corresponding 2D cross-section is higher than that of the 3D slope. This indicates that for slopes with large undulations, the 2D critical section does not accurately represent the 3D slope stability. Therefore, calculating the FoS of the full 3D slope is necessary. After 48 h of rainfall, the potential sliding surface in the 3D slope is located in the most prominent areas of the slope. Similarly, Fig. 6.17d, e presents the maximum shear strain and FoS for 3D slopes in Group 4 after 48 h of rainfall. When the undulation frequency is four, larger undulation amplitudes correspond to lower post-rain FoS values. The FoS of the 2D cross-section is lower than that of the 3D slope only when the undulation amplitude is 5 m. It further confirms that for large slope undulations, the 2D critical section

Table 6.3 Stability of 3D slopes with different undulation degrees and frequencies

Group	Model	Frequencies	Degrees (m)	FoS Rainfall duration (h)			DF (%)		
				0 h	24 h	48 h	0 h	24 h	48 h
4	O1	2	10	2.008	1.512	1.324	−2.58	−6.48	4.83
	O2	4	10	1.98	1.410	1.242	−4.04	−14.18	−1.45
	O3	6	10	2.008	1.367	1.180	−2.58	−17.78	−6.78
2D analysis				2.06	1.61	1.26			
5	O4	4	5	2.133	1.629	1.383	3.42	1.16	8.89
	O5	4	15	1.77	1.246	1.10	−16.4	−29.21	−14.54

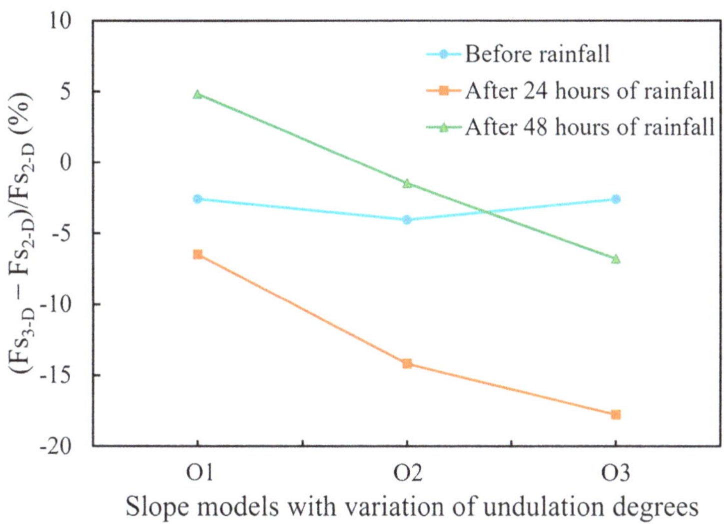

Fig. 6.18 Difference in the factor of safety for different frequencies of undulation (Li et al. 2024)

cannot reliably represent the 3D slope stability. Consequently, 3D FoS calculations are essential. As in Group 3, the potential sliding surface after 48 h of rainfall is concentrated in the most prominent areas of the slope. Table 6.3 summarizes the stability of 3D slopes with varying undulation amplitudes and frequencies, highlighting the influence of slope geometry on slope stability.

The results from Group 4 indicate that, for a constant undulation magnitude of 10 m, the factor of safety (FoS) of the 3D slope decreases as the undulation frequency increases. For frequencies of 4 and 6, the 3D slope FoS is lower than that of the selected 2D critical cross-section. This demonstrates that 2D analysis is not necessarily more conservative than 3D calculations for slopes with complex surface geometries. When the undulation frequency is held constant and the undulation magnitudes are increased to 10 m and 15 m, the 3D FoS remains lower than the 2D FoS. It indicates that larger undulation magnitudes amplify the discrepancy between 2D and 3D analyses. Figure 6.18 shows that prior to rainfall, the difference in FoS

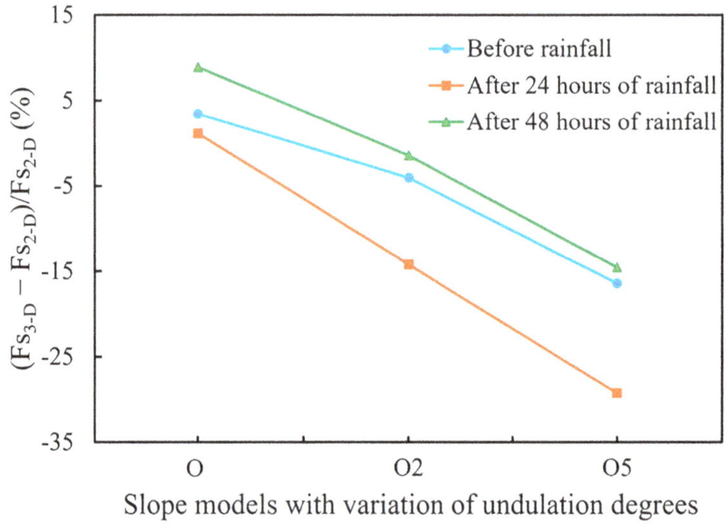

Fig. 6.19 Difference in the factor of safety for different degrees of undulation (m) (Li et al. 2024)

among 3D slopes with varying undulation frequencies is minimal when the undulation amplitude is 10 m. However, after 24 and 48 h of rainfall, the differences between 2D and 3D FoS become significant. Specifically, after 24 h of rainfall, the FoS differences between 3D slopes with undulation frequencies of 4 and 6 and the 2D critical section reach −14.18% and −17.78%, respectively. Overall, 3D slopes generally exhibit lower FoS than their 2D counterparts. Although the differences decrease after 48 h of rainfall, the FoS difference for the 3D slope with a frequency of 6 remains −6.78%. Figure 6.19 illustrates that when the undulation magnitude is small (e.g., 5 m), the 3D FoS exceeds that of the 2D cross-section. However, when the undulation magnitude reaches 15 m, the discrepancies become substantial. The FoS differences are −16.4%, −29.21%, and −14.54% before rainfall, after 24 h, and after 48 h of rainfall, respectively. These results clearly demonstrate that the traditional approach of selecting a 2D critical cross-section cannot replace a full 3D slope stability analysis. When slope surface undulations are significant, it can lead to substantial misjudgments in slope design.

6.5 Conclusions

To investigate how variations in spatial geometry affect slope stability under short-term heavy rainfall, four groups of three-dimensional slope models were designed for numerical simulation. The first group examines different increments in slope surface undulation while keeping the undulation frequency constant. Results indicate that larger undulation increments lead to lower slope stability after rainfall, even when the slope gradient remains constant. Specifically, when the undulation increment reaches 4 m along a given longitudinal length, the safety factors (FoS) of the 3D slope—before rainfall, after 24 h, and after 48 h of rainfall—are all lower than those of the selected 2D critical section. This demonstrates that 2D analysis is not necessarily more conservative than 3D analysis for slopes with pronounced geometric undulations. The second group introduces variations in slope gradient across the concave and convex sections of the slopes examined in Group 1. When the slope gradient varies significantly (from 40° to 64°), selecting a 2D critical section based on the maximum gradient produces overly conservative FoS estimates, potentially resulting in excessively cautious and uneconomical slope designs. The third group explores slopes with a fixed undulation magnitude of 10 m but varying undulation frequencies along the longitudinal length. The results show that higher undulation frequencies reduce the 3D slope FoS after short-term heavy rainfall. Even with a low undulation frequency of two, the 3D slope FoS before and after 24 h of rainfall is still lower than the corresponding 2D results. The fourth group considers a constant undulation frequency while varying the undulation magnitude. The results indicate that larger undulation magnitudes lead to lower 3D slope FoS. The discrepancy between 2D and 3D results becomes particularly pronounced when the undulation magnitude exceeds 10 m.

Overall, these findings highlight that complex 3D slope geometries can produce substantial differences in stability calculations compared to traditional 2D critical-section methods. Consequently, conducting full 3D slope stability analyses is essential for slopes with complex geometries. For reliable slope design, it is strongly recommended that geotechnical engineers perform both 2D and 3D analyses to capture the influence of spatial variations and ensure adequate safety.

References

Azizian A, Popescu R (2006) Three-dimensional seismic analysis of submarine slopes. Soil Dyn Earthq Eng 26:870–887. https://doi.org/10.1016/j.soildyn.2005.10.008

Dawson EM, Roth WH, Drescher A (1999) Slope stability analysis by strength reduction. Géotechnique 49:835–840. https://doi.org/10.1680/geot.1999.49.6.835

Griffiths DV, Marquez RM (2007) Three-dimensional slope stability analysis by elasto-plastic finite elements. Géotechnique 57:537–546. https://doi.org/10.1680/geot.2007.57.6.537

Jaboyedoff M, Carrea D, Derron M-H et al (2020) A review of methods used to estimate initial landslide failure surface depths and volumes. Eng Geol 267:105478. https://doi.org/10.1016/j.enggeo.2020.105478

Leong EC, Rahardjo H (2012) Two and three-dimensional slope stability reanalyses of Bukit Batok slope. Comput Geotech 42:81–88. https://doi.org/10.1016/j.compgeo.2012.01.001

Li AJ, Merifield RS, Lyamin AV (2010) Three-dimensional stability charts for slopes based on limit analysis methods. Can Geotech J 47:1316–1334. https://doi.org/10.1139/T10-030

Li X, Jiang Y, Sugimoto S (2024) Research on the influence of spatial dimensions on the stability of large-scale slopes under heavy rainfall. GeoHazards 5:732–754. https://doi.org/10.3390/geohazards5030037

Michalowski RL (2010) Limit analysis and stability charts for 3D slope failures. J Geotech Geoenviron Eng 136:583–593. https://doi.org/10.1061/(ASCE)GT.1943-5606.0000251

Nian T-K, Huang R-Q, Wan S-S, Chen G-Q (2012) Three-dimensional strength-reduction finite element analysis of slopes: geometric effects. Can Geotech J 49:574–588. https://doi.org/10.1139/t2012-014

Pan Y, Wu G, Zhao Z, He L (2020) Analysis of rock slope stability under rainfall conditions considering the water-induced weakening of rock. Comput Geotech 128:103806. https://doi.org/10.1016/j.compgeo.2020.103806

Rahardjo H, Ong TH, Rezaur RB, Leong EC (2007) Factors controlling instability of homogeneous soil slopes under rainfall. J Geotech Geoenviron Eng 133:1532–1543. https://doi.org/10.1061/(ASCE)1090-0241(2007)133:12(1532)

Sabzevari T, Noroozpour S (2014) Effects of hillslope geometry on surface and subsurface flows. Hydrgeol J 22:1593–1604. https://doi.org/10.1007/s10040-014-1149-6

Sun C, Chai J, Xu Z, Qin Y (2017) 3D stability charts for convex and concave slopes in plan view with homogeneous soil based on the strength-reduction method. Int J Geomech 17:06016034. https://doi.org/10.1061/(ASCE)GM.1943-5622.0000809

Talebi A, Troch PA, Uijlenhoet R (2008) A steady-state analytical slope stability model for complex hillslopes. Hydrol Process 22:546–553. https://doi.org/10.1002/hyp.6881

Tozato K, Dolojan NLJ, Touge Y et al (2022) Limit equilibrium method-based 3D slope stability analysis for wide area considering influence of rainfall. Eng Geol 308:106808. https://doi.org/10.1016/j.enggeo.2022.106808

Ugai K, Leshchinsky D (1995) Three-dimensional limit equilibrium and finite element analyses: a comparison of results. Soils Found 35:1–7. https://doi.org/10.3208/sandf.35.4_1

Wang J, Chen L (2021) The effect of hillslope geometry on Hortonian rainfall-infiltration-runoff processes. J Hydrol 594:125962. https://doi.org/10.1016/j.jhydrol.2021.125962

Wei WB, Cheng YM, Li L (2009) Three-dimensional slope failure analysis by the strength reduction and limit equilibrium methods. Comput Geotech 36:70–80. https://doi.org/10.1016/j.compgeo.2008.03.003

Wines D (2016) A comparison of slope stability analyses in two and three dimensions. J South Afr Inst Min Metall 116:399–406. https://doi.org/10.17159/2411-9717/2016/v116n5a5

Xu Y, Li J, Fan H et al (2017) Stability analysis of clastic rock slope with mudstone interlayer under rainfall infiltration. Geotech Geol Eng 35:1871–1883. https://doi.org/10.1007/s10706-017-0215-y

Xu J, Zhao X, Li P, Zhang M (2021) Stability of a 3D unsaturated vertical cut slope subjected to variable rainfall infiltration. Comput Geotech 134:104110. https://doi.org/10.1016/j.compgeo.2021.104110

Zhang Y, Chen G, Zheng L et al (2013) Effects of geometries on three-dimensional slope stability. Can Geotech J 50:233–249. https://doi.org/10.1139/cgj-2012-0279

Zhang LL, Fredlund MD, Fredlund DG et al (2015) The influence of the unsaturated soil zone on 2-D and 3-D slope stability analyses. Eng Geol 193:374–383. https://doi.org/10.1016/j.enggeo.2015.05.011

Zhang L, Gong W, Li X et al (2022) A comparison study between 2D and 3D slope stability analyses considering spatial soil variability. J Zhejiang Univ Sci A 23:208–224. https://doi.org/10.1631/jzus.A2100139

Zheng H (2012) A three-dimensional rigorous method for stability analysis of landslides. Eng Geol 145–146:30–40. https://doi.org/10.1016/j.enggeo.2012.06.010

Zhou Y, Qi S-C, Fan G et al (2020) Topographic effects on three-dimensional slope stability for fluctuating water conditions using numerical analysis. Water 12:615. https://doi.org/10.3390/w12020615

Zienkiewicz OC, Humpheson C, Lewis RW (1975) Associated and non-associated visco-plasticity and plasticity in soil mechanics. Géotechnique 25:671–689. https://doi.org/10.1680/geot.1975.25.4.671

Chapter 7
Evaluation of Three-Dimensional Slope Stability Under Heavy Rainfall Based on Geometric Features

Abstract Building upon the methods and findings of previous chapters, this chapter proposes a quantitative approach for describing slope geometric features. Inspired by the concept of rock joint roughness, a Slope Surface Roughness Coefficient (SSRC) is introduced to characterize the variability of slope surfaces and gradients. The method is applied to a case study of a rocky slope along National Route 251, spanning approximately 4500 m along the coastline from Unzen City to Minamishimabara City, Nagasaki, Japan. Slope gradients and surface undulations of 3D slopes were extracted using both Rhino and ArcGIS, and the SSRC values were calculated accordingly. Based on the linear relationship between SSRC and the Factor of Safety (FoS), the approach was used to assess safety factor variations for six actual slopes measuring 140 m × 140 m with heights of approximately 80 m under rainfall conditions. The predicted values closely match numerical simulation results, demonstrating that the SSRC-based method provides an efficient and accurate means of estimating 3D slope stability under rainfall. This approach not only offers a practical tool for individual slope assessment but also establishes a solid foundation for subsequent regional-scale slope stability evaluations using machine learning models, facilitating more informed decision-making in geotechnical engineering and hazard mitigation.

Keywords Complex slope geometry · SSRC · Factor of safety · Maximum shear strain · Potential failure zone

7.1 Introduction

The assessment of regional landslide susceptibility under heavy rainfall is crucial for effectively preventing geological hazards that arise from slope failures. Methods for assessing rainfall-induced landslide susceptibility are primarily categorized into qualitative and quantitative approaches (Park et al. 2013). Qualitative methods rely

Y. Jiang, X. Li, *Stability Evaluation of Slope Under Heavy Rainfall Based on Geometric Features*, Springer Natural Hazards,
https://doi.org/10.1007/978-981-95-8656-1_7

heavily on expert judgment, making them less suitable for large-scale assessments due to their lack of a solid physical basis in slope failure mechanics. The quantitative methods can be further divided into statistical methods and geotechnical approaches. The geotechnical approach focuses on analyzing the mechanical conditions of slopes and assessing their stability through mathematical calculations, providing a more rigorous framework for evaluating landslide susceptibility (Lee and Chi 2011). In this approach, the physical properties of a particular slope are obtained from field investigating and laboratory tests, and a physical based model is used to evaluate landslide susceptibility. These physically based methods do not rely on historical data. Instead, they simulate real-world physical processes to make predictions, which makes them particularly effective in systems where the underlying mechanisms are well understood (Gutiérrez-Martín 2020).

The occurrence of rainfall-induced landslides is influenced by a variety of spatial and climatic factors, including geological composition, topography and hydrogeological conditions (Park et al. 2013). Consequently, Geographic Information Systems (GIS) have become vital for managing and analyzing spatial information related to landslides, given their robust capabilities in processing distributed data effectively across multiple dimensions (Van Westen et al. 2006). Most existing GIS-based methods typically rely on a one-dimensional (1D) infinite slope model (Gutiérrez-Martín 2020) or a two-dimensional (2D) slope stability model that uses cross-sections of a digital elevation model (DEM) (Gu et al. 2015). Although 1D and 2D approaches are often preferred for their simplicity and perceived conservative estimates, they tend to underestimate slope stability when compared to 3D methods (Tran et al. 2017). When specific combinations of soil and groundwater characteristics, complex geometries (e.g., ridges and corners), and boundary conditions are considered in the numerical simulation, it is highly recommended to conduct the 3D slope stability analysis instead of only calculating the 2D critical cross-sections (Li et al. 2024). However, three-dimensional (3D) slope stability assessment at a regional level is not commonly practiced due to its complexity, high computational demands, and lack of sufficient data (Tran et al. 2017). Therefore, making necessary assumptions is crucial for simplifying the calculation process in geotechnical analysis, enabling more efficient handling of complex analyses while still maintaining accuracy in the results (Van Den Bout et al. 2021). Most studies on regional rainfall-induced landslide susceptibility, whether employing GIS or other methods, rely on various assumptions to simplify the calculations (Xie et al. 2004; Ho et al. 2012; Park et al. 2013; Jia et al. 2015; Tran et al. 2022).

In general, regional slope stability analysis under rainfall conditions typically consists of two main components: infiltration analysis and slope stability evaluation. Commonly used infiltration analysis methods is the Green-Ampt (GA) model (Heber Green and Ampt 1911), a simple model that describes one-dimensional infiltration behavior. This model is widely favored in many studies due to its low computational cost, making it particularly suitable for large-scale landslide susceptibility assessments. Tozato et al. (2022) developed a comprehensive method for assessing the risks associated with slope failures caused by heavy rainfall across extensive areas. For the infiltration analysis, the Green-Ampt model was utilized to estimate the depth of water infiltration under rainfall conditions. The depth of infiltration derived from the infiltration analysis serves as critical input for the slope stability evaluation, which links the hydrological processes to stability assessments. However, the GA model assumes an idealized scenario for rainfall infiltration into unsaturated soils, overlooking real-time variations in soil permeability and matric suction during the infiltration process. These factors are crucial for accurately assessing slope stability under rainfall conditions. In other words, it does not capture the soil-water characteristic curve (SWCC) of the unsaturated soil, which plays a key role in describing the relationship between soil suction and water content. The neglect of significant impact of matric suction on the shear modulus of soils (Hoyos et al. 2015) further undermines the accuracy of the computed factor of safety, potentially leading to less reliable slope stability assessments. This limitation emphasizes the need for more advanced models that account for dynamic changes in soil properties during infiltration for enhanced precision in slope stability analysis.

Given the difficulties in accurately measuring spatial parameters such as initial groundwater conditions across large areas, this study opted for a steady-state model over a transient model to simplify the analysis (Park et al. 2013). The steady-state hydrogeological model assumes that rainfall infiltration occurs at a steady state, with saturated water flowing parallel to the slope surface. Tran et al. (2017, 2022) utilized the Transient Rainfall Infiltration and Grid-Based Regional Slope-Stability (TRIGRS) model (Baum et al. 2010) to predict potential landslides in a three-dimensional context. These models were employed to simulate the effects of dynamic pore pressure changes induced by rainfall, enabling a more accurate assessment of slope stability under varying hydrological conditions (Palazzolo et al. 2021). In their study, the rainfall infiltration and vertical flow fluxes through the unsaturated zone are also described in TRIGRS using the Richards equation. This equation offers a robust representation of the dynamic processes governing water movement in unsaturated soils, demonstrating notable improvements over the GA model (Iverson 2000). However, this approach requires a substantial amount of input data, including digital elevation models (DEMs), geotechnical and hydraulic properties, initial soil moisture levels, and soil depth for each individual cell (Tran et al. 2017). Consequently, the effects of negative pore water pressure were excluded from the calculations due to the limitations in available data for simulation, with all negative values set to zero in their study. Current research on the assessment of regional rainfall-induced landslides predominantly focuses on the stability analysis of rainfall infiltration in weathered soils (Xie et al. 2004; Van Westen et al. 2006; Gutiérrez-Martín 2020). However, if the permeability of the rock layer is relatively high, rainfall can infiltrate into the weathered rock layer even during short-term rainfall events, resulting in a further decrease in the slope stability. Unlike the shear strength of the weathered soil layer, which is mainly influenced by matric suction, the physical and mechanical parameters of the rock layer are significantly affected by water-induced weakening effects (Cai et al. 2020). For the same rock type, the reduction coefficient varies due to differences in rock porosity, typically exhibiting an exponential relationship between the reduction coefficient (λ) and porosity (Li et al. 2012; Pan et al. 2020). Accurate calculation of unsaturated seepage of rainwater in both soil and rock layers is critical for evaluating the slope stability under rainfall conditions.

On the other hand, the assumptions used to simplify three-dimensional slope stability calculations are primarily based on the limit equilibrium method (LEM), which is widely adopted in many GIS-based studies. The 3D LEM is known for being both simple and efficient (Zhang et al. 2013b), typically achieved by extending various 2D slice methods into 3D column methods, with several enhancements (Hovland 1977; Zheng 2012). For instance, the extended Bishop simplified method of slices for limit equilibrium methods (Bishop 1955) provides a straightforward and efficient approach applicable to a variety of practical problems. This method assumes that the lateral forces acting on soil columns are horizontal, effectively eliminating net shear stress between slices. The factor of safety (FS_{3D}) for three-dimensional stability analysis is determined by summing the contributions from all columns along the potential slip surface (Tran et al. 2017). In their study, slope stability analysis was performed by the limit equilibrium method using the Hovland method (Hovland 1977), a relatively simple approach to three-dimensional slope stability analysis, was applied and assessed (Jia et al. 2015). While this method is computationally efficient for regional-scale analyses of slope stability under rainfall conditions, an overestimation of the factor of safety (FoS) was observed (Jia et al. 2012). A column-based 3D limit equilibrium model revised Hovland model was also used to calculate the safety factor of each slope unit (Xie et al. 2007). They adopted a Monte-Carlo simulation to locate possible slip surfaces and then applied the revised Hovland model to calculate the corresponding safety factors. It assumed the slip surface of the 3D slope to be the lower part of an ellipsoid. Such assumptions inherently simplify the analysis, as the complex slope geometry often results in a failure surface that deviates significantly from an ellipsoidal shape. They have pointed out that a significant drawback of 3D LEM is the challenge of accurately identifying the critical 3D failure surface, both in terms of its shape and location (Wei et al. 2009). Consequently, the random search process across the entire study area, based on the assumption of an ellipsoidal slip surface, is likely to introduce errors in determining the volume and geometry of potential landslides (Van Den Bout et al. 2021). As concluded in their study, the method of simulating an ellipsoid-shaped initial slip surface may overlook certain potential slip surfaces (Xie et al. 2004).

Unlike existing GIS-based methods that assess the reginal slope stability relying on simplified models like the GA model or the one-dimensional Richard equation. This study uses FLAC^{3D} to perform a detailed analysis of unsaturated infiltration in a single 3D slope model. The shear strength reduction method (SSRM) is applied to calculate the variation in the safety factor of 3D slopes during rainfall infiltration (Zienkiewicz et al. 1975). The changes in pore water pressure in unsaturated soils can be described by incorporating matric suction through Bishop's effective stress principle (Ng and Shi 1998; Cai and Ugai 2004). For unsaturated seepage in rock layers due to further rainfall infiltration, the physical and mechanical parameters of the slope model are dynamically adjusted in real time based on changes in saturation levels (Pan et al. 2020). The deterministic calculations based on the geotechnical approach aim to accurately reflect the variations in the factor of safety (FOS) of three-dimensional slopes under rainfall conditions. The objective of this study is to assess regional landslide susceptibility under rainfall conditions by integrating the computational results from $FLAC^{3D}$ with analytical data from GIS. This study first utilizes GIS to identify potential landslide hazard zones and then conducts detailed analyses for specific areas. The variation in the factor of safety (FOS) for a 3D slope under rainfall conditions is influenced by several interrelated factors, primarily soil shear strength parameters such as cohesion c and internal friction angle φ, slope geometry and soil thickness (Ho et al. 2012). As more data is collected, the scope of the regional landslide hazard assessment can be gradually expanded based on these factors of the three-dimensional slopes. This work lays a solid foundation for using machine learning methods in future studies to predict rainfall-induced landslides on a regional scale after accumulating a large amount of data.

7.2 Methodology

This study employs a deterministic method to calculate the variations in the safety factor (FoS) of three-dimensional slopes under short-term heavy rainfall. The Bishop's effective stress principle based on the Mohr-Coulomb criterion is employed to describe the changes in matrix suction of unsaturated soil during the rainfall infiltration (Ng and Shi 1998; Cai and Ugai 2004). For situations where rainwater further infiltrates into the rock layers, the water-induced weakening of rocks is used to characterize the variations in the physical and mechanical parameters of the rock in response to water infiltration (Vásárhelyi and Ván 2006; Li et al. 2012; Iyare et al. 2023). An empirical reduction factor λ of 0.8 is used in the numerical simulations when the rock layer of the slope is in a saturated state (Xu et al. 2017). The calculation of unsaturated seepage in slopes has been thoroughly explained in the previous sections, so it is not elaborated in this chapter. To quantify the relationship between the geometric characteristics of 3D slopes (slope gradient and surface undulation) and the factor of safety (FoS) under rainfall conditions, this section introduces the concept of the Slope Surface Roughness Coefficient (SSRC). Four groups of standard slope models were designed to analyze the correlation between the SSRC and the FoS, as shown in Figs. 6.2 and 6.6. For a detailed description of the calculation and analysis process for these 3D standardized slope models, please refer to our previously published research (Li et al. 2024). The target site is a rocky slope along National Route 251, extending approximately 4500 m along the coastline from Unzen City to Minamishimabara City in Nagasaki, Japan. To assess the safety of the steep slopes along the national route, detailed investigations were conducted to understand the ground properties of the area. These investigations included borehole drilling, standard penetration tests, in situ tests, and triaxial compression tests. The geotechnical parameters for the slope model are presented in Table 6.1. All the numerical simulations were conducted under continuous rainfall conditions with a constant intensity of 20 mm/h. To ensure consistency across all models, other variables affecting the slope safety factor are held constant in the calculation. Specifically, the thickness of the two layers of weathered rock is set at 3 m, the slope height is 80 m, and the groundwater level is positioned relatively low based on the field investigation (Fig. 6.4b). To ensure controlled variables during the numerical simulation process, the physical and mechanical parameters of the slopes in this chapter's calculations are kept consistent.

The Slope Surface Roughness Coefficient (SSRC) quantifies the geometric characteristics of the slope surface and angle, specifically measuring the degree and frequency of surface undulation. As shown in Fig. 6.5a, the dashed line represents the reference plane for quantifying the surface undulation of the slope, where the convex portions signify positive undulation, and the concave portions denote negative undulation. The impact of individual concave and convex-shaped slopes, as well as turning corners, on slope stability has been extensively studied (Nian et al. 2012; Zhang et al. 2013a; Zhou et al. 2020). The conclusion indicates that concave slopes yield a factor of safety (FoS) greater than that of typical slopes, while convex slopes produce a lower FoS (Zhang et al. 2015). In the case of concave slopes, the concavity can create increased lateral confinement for the soil, resulting in higher safety factors in three-dimensional analyses. In contrast, convex slopes tend to concentrate stress, particularly near the crest, which raises the risk of local sliding. As a result, three-dimensional analyses of convex slopes may yield lower safety factors. These slopes with continuous undulations exhibit greater complexity in their safety factor assessments compared to regular slopes which are relatively smooth and have simpler geometries. To facilitate comparison between the SSRC of different three-dimensional slope models, the calculation of SSRC only considers the adverse effects of the convex portions of continuously undulating slopes on the factor of safety (FoS), while the data for the concave portions is excluded from the SSRC calculation.

The slope model has a longitudinal length of 140 m, divided into 28 two-dimensional cross-sections, with each section spaced 5 m apart (Fig. 7.1a). The slope angle and surface undulation were measured for each cross-section. Since the slope surface remains unchanged along the direction perpendicular to the longitudinal axis (Fig. 6.3b), the maximum and average values for both slope angle and surface undulation are the same (Fig. 7.1a, b). As shown in Fig. 7.1d, when the slope angle is constant (50°), the maximum convexity of the slope surface in Model P4 is 13 m, and the area of maximum shear strain (potential slip area) occurs within the most convex region of the slope. As shown in Fig. 7.2d, although the maximum convexity of the slope surface in Model P2.1 is 9 m, the area of maximum shear strain (potential slip area) does not occur at the most convex part of the slope. When the slope angle is steep (64°), the potential slip area is concentrated in the region with the steepest gradient. This demonstrates the significant influence of slope angle on the factor of safety (FoS) for the 3D slope. As a result, when assigning weights in the SSRC calculation, the slope angle is given a higher weighting than surface undulation. This is further supported by the FoS results for Group 2 in Table 6.2. The FoS values for Models P1.1, P2.1 and P3.1 which account for the variation of

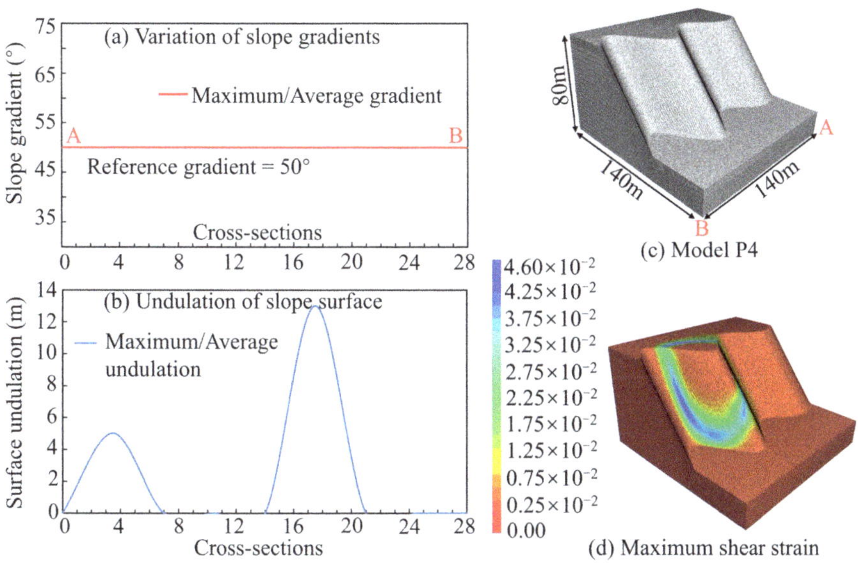

Fig. 7.1 The statistical data for slope angle (**a**) and surface undulation (**b**) of Model P4 (**c**) with the maximum shear strain (**d**) before rainfall

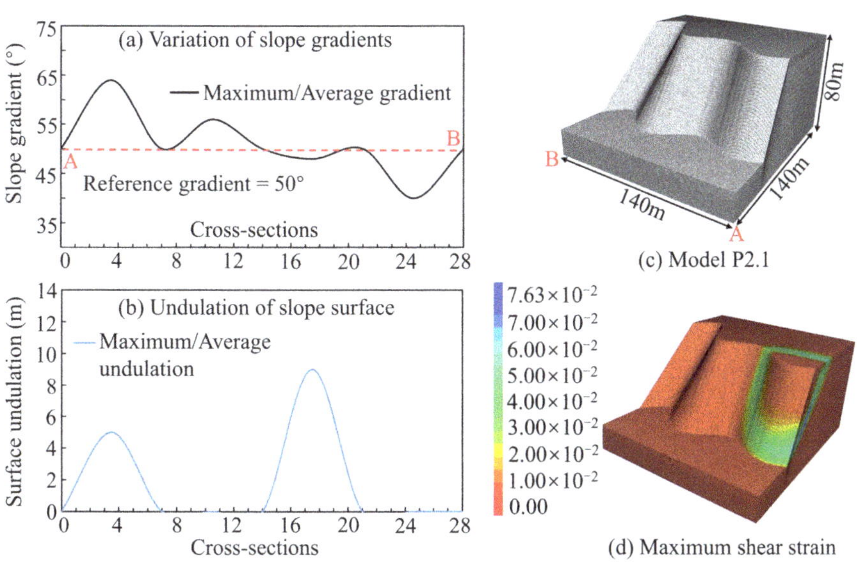

Fig. 7.2 The statistical data for slope angle (**a**) and surface undulation (**b**) of Model P2.1 (**c**) with the maximum shear strain (**d**) before rainfall

slope gradient, are consistently lower than that of Model P4 which has the highest surface undulation. Apart from the degree of undulation, another significant geometric characteristic affecting the FoS is the frequency of undulation, or the number of turning corners over a certain length in the longitudinal direction. As shown in Table 6.3 for Group 3, when the undulation frequency increases from 2 to 6, the post-rainfall FoS decreases significantly with the rise in undulation frequency. The results from Group 4 indicate that when the degree of undulation is sufficiently high (e.g. 15 m), the FoS of the slope decreases significantly. Compared to the model with an undulation degree of 10 m and a frequency of 6 (Model O3), the safety factor is notably lower for Model O5 with an undulation of 15 m. This suggests that when the degree of undulation exceeds 10 m, its weight in the SSRC calculation becomes more significant than that of the surface undulation frequency.

7.2.1 Slope Surface Roughness Coefficient (SSRC)

The analysis reveals that the geometric characteristics affecting the FoS of 3D slopes include slope gradient, degree and frequency of surface undulation. Therefore, the Slope Surface Roughness Coefficient (SSRC) calculation incorporates these factors, assigning different weights to each based on their respective influence on the FoS. The equation of calculating SSRC is given as follows:

$$SSRC = w_1 \cdot e^{\left(\frac{\Delta\theta_{\max}}{\Delta\theta_{\max}^{Ref}}\right)} + w_2 \cdot e^{\left(\frac{\Delta\theta_{avg}}{\Delta\theta_{avg}^{Ref}}\right)} + w_3 \cdot \frac{\Delta h_{\max}}{\Delta h_{\max}^{Ref}} + w_4 \cdot \frac{\Delta h_{avg}}{\Delta h_{avg}^{Ref}} + w_5 \cdot \frac{f_q}{f_q^{Ref}} \quad (7.1)$$

where $\Delta\theta_{\max}$ represents the maximum slope gradient of the actual slope, while $\Delta\theta_{\max}^{Ref}$ denotes the maximum slope gradient of the reference model. $\Delta\theta_{avg}$ is the average slope gradient of the actual slope, and $\Delta\theta_{avg}^{Ref}$ is the average slope gradient of the reference slope model. $\Delta h_{\max}$ indicates the maximum surface undulation of the actual slope, and $\Delta h_{\max}^{Ref}$ refers to the maximum surface undulation of the reference slope model. Δh_{avg} is the average surface undulation of the actual slope, while Δh_{avg}^{Ref} is the average surface undulation of the reference slope model. f_q represents the frequency of surface undulation exceeding a certain threshold (10 m) in the longitudinal direction of the actual slope model. f_q^{Ref} is the number of surface undulation exceeding 10 m of standard slope model. The weights w_1, w_2, w_3, w_4 and w_5 correspond to the coefficients for each indicator in the equation.

It is worth noting that the maximum slope gradient in the Eq. (7.1) refers to the maximum value among the 28 two-dimensional cross-sections, and the average slope gradient refers to the average value across these 28 cross-sections. Table 6.2 indicates that when the surface undulation is 5 m, the factor of safety (FoS) calculated for the 3D model does not decrease compared to the 2D analysis. Thus, the threshold for frequency f_q is defined as the occurrence of undulation exceeding or equal to 10 m. The Model O4 is used as the reference slope model for calculating the

SSRC of models in Group 2. In comparison to the results of the 2D analysis, Model 4, which has an undulation degree of 5 m and a frequency of 4, exhibited an increase in the safety factor rather than a decrease. This suggests that when the geometric characteristics of the slope are less pronounced, the safety factor of the 3D slope is greater than that of the corresponding 2D slope. This phenomenon can be attributed to the end effects inherent in 3D slope stability analysis (Zhang et al. 2013b). Since models in Group 2 are more in line with the actual slope with varying undulations and slope gradients, they were used to analyze the relationship between SSRC and FoS. A manual analysis method is currently employed to determine the values of weights. For now, the weights are set as follows: $w_1 = 0.6$, $w_2 = 0.25$, $w_3 = 0.05$, $w_4 = 0.05$, $w_5 = 0.05$. Due to the limited amount of data, the weighting coefficients and the calculation of the SSRC in this section are primarily based on the results from Tables 6.2 and 6.3, as well as the authors' practical experience in engineering. Once sufficient data has been accumulated in the future, more accurate selections of the weighting coefficients and SSRC calculations can be conducted. In this formulation, it is common practice for the total weights (w_1, w_2, w_3, w_4 and w_5) to sum to 1. This normalization allows the weights to be viewed as proportions, reflecting the relative importance of each factor in contributing to the overall surface roughness.

7.2.2 Relationship Between SSRC and FoS

Table 7.1 presents the results of SSRC for Models P1.1 to P4.1. The data shows that as the SSRC value increases, the geometric characteristics of the three-dimensional slope that contribute to a lower safety factor become more prominent, resulting in a smaller safety factor after the same duration of rainfall. This demonstrates that the formulation of the SSRC and the choice of weighting factors are well-founded. Inspired by Barton's approach of using the Joint Roughness Coefficient (JRC) to estimate the shear strength parameters of rock (Barton and Choubey 1977), this study proposes the use of the Slope Surface Roughness Coefficient (SSRC) to calculate the safety factor of three-dimensional slopes. Given the current limited data, a linear relationship is temporarily used to fit the correlation between SSRC and FoS, as illustrated in Fig. 7.3.

Table 7.1 SSRC of models in Group 2

		FoS		
		Rainfall duration (h)		
Model	SSRC	0	24	48
P1.1	3.00	1.78	1.340	1.195
P2.1	3.06	1.77	1.33	1.19
P3.1	2.88	1.934	1.461	1.285
P4.1	2.62	2.09	1.527	1.332
O4 (standard)	2.46	2.133	1.629	1.383

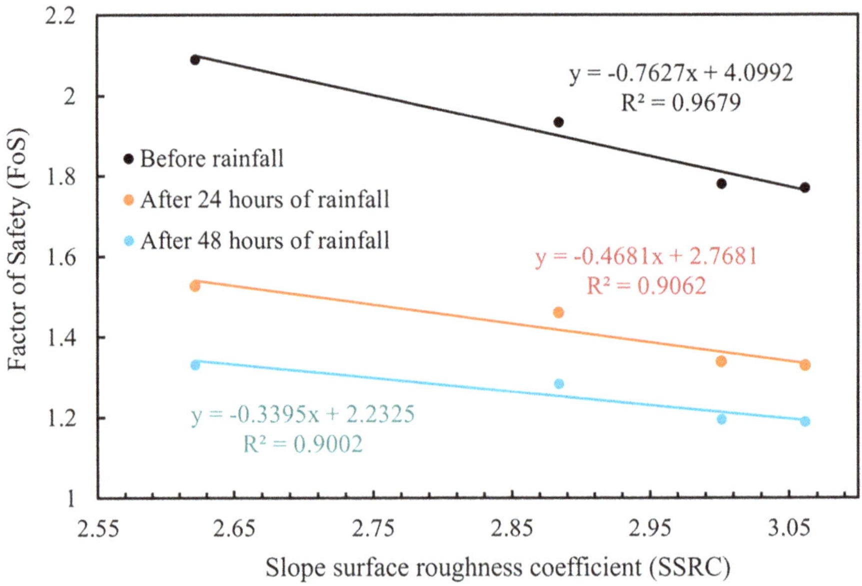

Fig. 7.3 The fitted relationship between SSRC and FoS

Fig. 7.4 The photograph of the slope in Region 1 (**a**) and Region 2 (**b**) along the coastal road

7.3 Evaluation of 3D FoS Based on SSRC

The target site is a rocky slope along National Route 251, extending approximately 4500 m along the coastline from Unzen City to Minamishimabara City in Nagasaki, Japan (Fig. 6.1a). Figure 6.1c, d illustrate two slope models extracted from Regions 1 and 2 within the study area, while Fig. 7.4 displays two photos of steep and high slopes located in Regions 1 and 2, respectively. To thoroughly assess the safety of these steep slopes along the national route, comprehensive investigations were conducted to evaluate the geological properties of the area. This included borehole drilling, standard penetration tests, in situ tests, and triaxial compression tests. The findings from the geological survey indicate that the thickness of the weathering zone in the rock slope within the target area ranges from approximately 6 to 8 m. The surface layer is characterized as a mixture of colluvial soil and heavily weathered sandstone, with the thickness of the first weathered layer set at 3 m for this analysis. The second layer consists of moderately weathered sandstone and mudstone, also with a thickness of 3 m. Beneath these layers lies the bedrock, primarily composed of fresh sandstone. Table 6.1 outlines the geotechnical properties of the slopes in saturated conditions, revealing that the slopes in this area are predominantly made up of soft rock.

This section introduces a method that combines GIS and FLAC^{3D} to assess the safety factor of slopes in a specified area based on SSRC under rainfall conditions. The assumptions of this study should be clearly outlined before the calculations are presented. Field investigations indicate that groundwater levels in the study area are generally low. Given the focus on shallow slope failure of this study, it is reasonable to analyze the impact of short-term heavy rainfall without considering changes in groundwater levels (Ho et al. 2012). Since the heights of the slopes along the coastline mainly range from 60 to 70 m, the influence of minimal variation in slope height was not considered in the calculation. The shear strength parameters (cohesion c and internal friction angle φ) of the slope materials within a specific area were assumed to be uniform. It should be noted that soil thickness is a key factor in determining the regional assessment of susceptibility to rainfall-induced shallow landslides (Xie et al. 2004; Tran et al. 2017). Soil depth distribution results from a complex interplay of factors, including topography, parent material, climate, and various biological, chemical, and physical processes. However, the relationship between regional soil depth and the safety factor of three-dimensional slopes requires further computational data for support. For now, soil depth is assumed to be uniformly distributed with a thickness of 3.0 m (Segoni et al. 2012).

7.3.1 Extract Topographic Information in Rhino

Figure 7.5 shows a flowchart of the process for extracting SSRC data of three-dimensional slopes using Rhino software. Firstly, ArcGIS was used to determine geometric characteristics of the slope (slope gradient and height) in a specific area, allowing for a preliminary assessment of landslide hazard zones (Fig. 7.6). It can be seen that the slope gradient of area along the coastline is primarily around 50°, which falls within a high slope gradient range and requires further stability analysis. Figure 7.7 shows the statistical data of slope angle and height based on the divided slope units. These two geometric characteristics of slope units are important factors influencing slope stability. However, the information provided in Figs. 7.6 and 7.7 can only offer a general indication of potential landslide hazard areas along the coastal highway slopes. A more accurate analysis of 3D slope stability under heavy rainfall requires more detailed calculations in FLAC^{3D}.

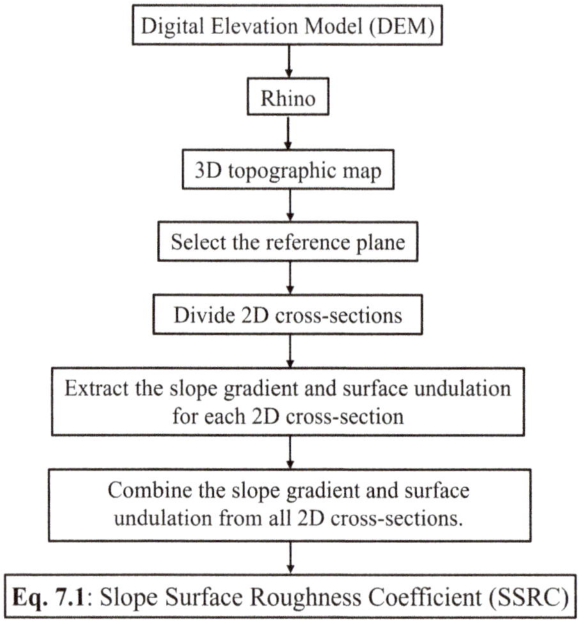

Fig. 7.5 The flowchart of the process for extracting SSRC data in Rhino

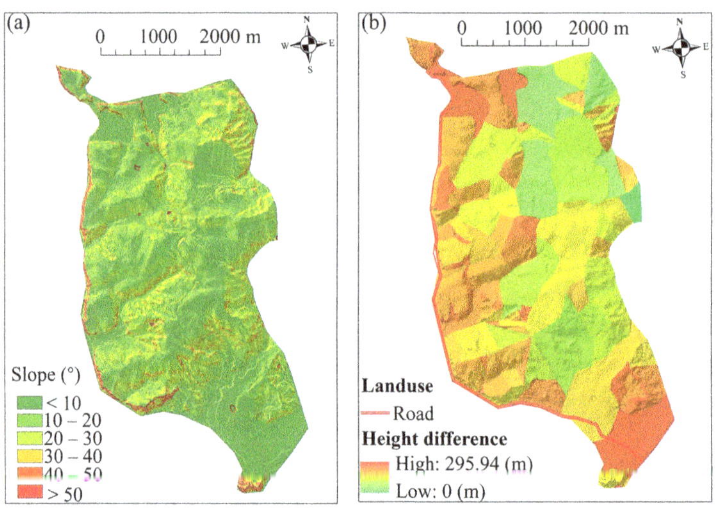

Fig. 7.6 (**a**) The slope gradient of the study area and (**b**) the height difference of the study area

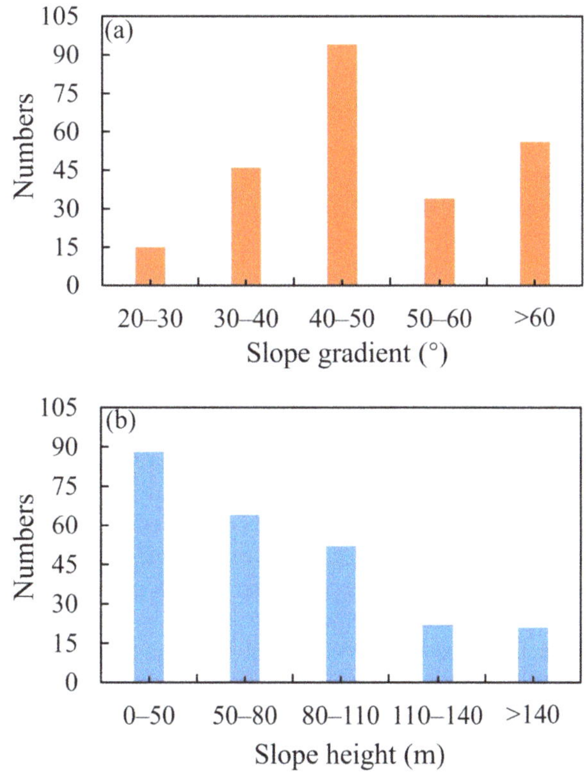

Fig. 7.7 (**a**) The slope gradient statistics and (**b**) the slope height statistics of the study area

The height and slope angle of a slope are the two most fundamental and crucial factors determining its stability. These two parameters play a key role in understanding slope behavior under various conditions. They influence the forces acting on the slope, the material properties, and external factors such as rainfall, seismic activity, and human-induced disturbances. The relationship between the height and slope angle can directly influence the potential for slope failure. Therefore, it is essential to assess these factors comprehensively when performing slope stability analyses. First and foremost, the height of the slope is an important factor because it determines the magnitude of the gravitational forces acting on the slope material. The higher the slope, the greater the mass of material at risk of movement, which increases the potential for failure. Additionally, the height of a slope influences the depth to which water can infiltrate during rainfall, which can increase pore water pressure and reduce the effective stress in the soil, making it more prone to failure. When combined with a steep slope, this increased mass and reduced strength due to water saturation can lead to catastrophic slope failure, such as landslides or rockfalls. The slope angle, or the steepness of the slope, is equally critical. As the slope angle increases, the driving forces that cause the soil or rock to move, such as the component of gravitational force acting parallel to the slope, also increase. In general, as the slope angle approaches or exceeds the critical angle of repose (the maximum angle at which loose material can be piled without sliding), the stability of the slope significantly decreases. A steeper slope is more vulnerable to failure because the shear stresses within the slope material become larger, potentially exceeding the shear strength of materials.

Furthermore, the steeper the slope, the more likely it is to experience rapid and large-scale movements, which are often difficult to predict and manage. The interaction between slope height and slope angle can exacerbate the stability issue. For example, a tall and steep slope may have a high factor of safety under dry conditions. But the pore water pressure can rise dramatically during heavy rainfall or snowmelt, reducing the effective stress in the soil. This can weaken the slope further and lead to a collapse. Additionally, the specific material properties of the slope, such as cohesion, friction angle, and permeability, can vary with height and slope angle. This further complicates the analysis of slope stability. In practical applications, understanding the relationship between slope height and angle is essential for designing and maintaining safe slopes. Engineers and geologists use a variety of tools, including slope stability models, to analyze how different combinations of height and slope angle affect the overall stability of a slope. These models typically take into account the type of soil or rock, the presence of vegetation, external factors such as seismic activity, and water content in the material. By thoroughly assessing the height and slope angle, along with other factors, more effective mitigation strategies can be developed to prevent slope failures. In conclusion, the height and slope angle of a slope are integral to its stability, and understanding their effects is crucial in geotechnical engineering. These factors interact with each other and with other environmental conditions to determine whether a slope is likely to fail. Consequently, accurate assessment and consideration of these parameters are fundamental to ensuring the safety and stability of slopes, particularly in regions vulnerable to natural hazards.

The two regions highlighted by red boxes in Fig. 6.1b, Region 1 and Region 2, were selected as case studies for comparative analysis. These regions were used to examine how slope geometric characteristics, such as slope gradient and surface undulation, affect the safety factor under rainfall conditions. The main difference between these two regions lies in the degree of undulation on the slope surface. GIS was used to extract contour lines of Region 1, which were then imported into Rhino to create a three-dimensional topographic map for further calculation of the SSRC, as shown in Fig. 7.8. A critical factor in the calculation of SSRC is the selection of the reference surface for the surface undulation of three-dimensional slopes. This aspect is much easier to achieve using Rhino software compared to GIS. Additionally, this software can effortlessly generate two-dimensional cross-sections based on user requirements, as illustrated in Fig. 7.8a. Given that the DEM elevation map has a resolution of 5 m × 5 m, a two-dimensional cross-section was extracted every 5 m along the 420-m length of Region 1 to collect geometric information about the 3D slope. This area was subsequently divided into three smaller areas, namely Areas 1, 2 and 3 for further calculations of slope models in FLAC^{3D}. Each area has a length of 140 m and a width of 200 m, and it contains 28 two-dimensional cross-sections.

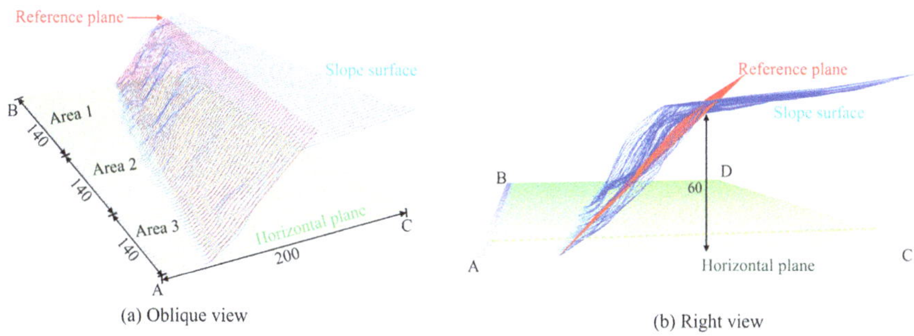

Fig. 7.8 Three-dimensional topographic map of Region 1 and the generated cross-sections (unit: m)

A reference plane must be established to calculate the degree of concave or convex undulation. The method for determining the reference plane involves selecting the two-dimensional sections at the start (AC) and end (BD) cross-sections (Fig. 7.8b). As shown in Fig. 7.9a, for the two-dimensional cross-section AC, point O was selected at the slope toe and point R was selected at the crest. A straight line (OR) was drawn to connect these two points, defining the gradient of the reference line for this cross-section. To improve the accuracy of the statistical analysis, this reference line (OR) was designed to segment the undulations of the two-dimensional cross-section as evenly as possible. This approach minimizes the discrepancy between the actual two-dimensional slope profile and the standard slope represented by the reference line. A similar process was applied to the two-dimensional cross-section BD, where another reference line was determined. These two reference lines for the initial cross-section 3.1 and the final cross-section 3.28 were designed to maintain a consistent slope angle of 43°, as shown in Fig. 7.9a, b. The two reference lines were then connected to form a plane, serving as the reference plane for evaluating slope surface undulation (as shown in Fig. 7.8). Once this reference plane was established, it results in four typical two-dimensional cross-sections within Regions 1 and 2, as illustrated in Fig. 7.9.

Figure 7.9d represents one of the two-dimensional cross-sections within Region 2. It can be observed that the standard slope in this region, defined by the reference plane, forms a gradient of 52° with the horizontal plane. This is different from the standard slope in Region 1, where the standard gradient is 43°. Although manual control can create an ideal segmentation between the reference plane and the slope surface for the cross-sections (Fig. 7.9a, c), other cross-sections may exhibit the conditions shown in Fig. 7.9b, d. This highlights the necessity of the SSRC (Slope Surface Roughness Coefficient) proposed in this study, which employs a standardized method to describe the degree of undulation in 3D slopes along the longitudinal direction. Due to the differences of selection between the first and last two-dimensional cross-sections, the reference planes for Region 1 and Region 2 have different slope gradients, 43° and 52° respectively. Therefore, when comparing the SSRC of the 3D slopes in subsequent analysis with the SSRC of standard Models

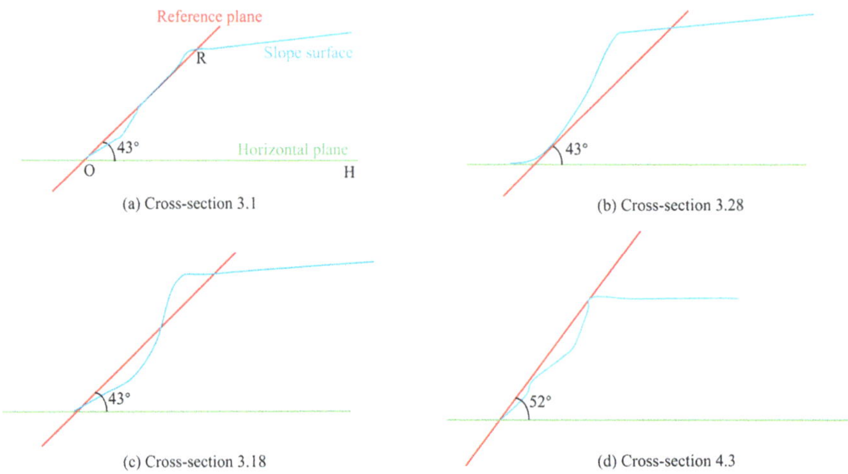

Fig. 7.9 Four typical types of two-dimensional cross-sections. The red line represents reference plane, the blue line is the slope surface, and the green line is the horizontal plane

P1 to O5, which have a gradient of 50°, a reduction coefficient λ will be applied to ensure the accuracy of the FoS predictions.

Based on the impact of slope gradients on the calculation of the three-dimensional slope safety factor and the errors encountered during the SSRC calculation, the reduction factor λ_1 of SSRC values for the first set of slope models (Slope models 1, 2, and 3) were set to 1.6, while Slope models 4, 5 and 6 were adjusted using the factor $\lambda_2 = 1.1$. When sufficient data on the SSRC and safety factors of three-dimensional slopes are accumulated, such correction factors λ can be supported by more robust data. Currently, however, the determination of these factors largely depends on the practical engineering experience of engineers. This approach, while effective in the short term, can be subjective and may not always account for the full complexity of slope behavior. As more data becomes available, it will be possible to transition towards a more data-driven methodology.

7.3.2 *Extract SSRC Data in ArcGIS*

Figure 7.10 shows a flowchart of the process for extracting SSRC data of three-dimensional slopes using ArcGIS software. Similar to the method used in the previous section, this section primarily utilizes the powerful capabilities of ArcGIS for extracting topographic information to obtain slope gradient and surface undulation for three-dimensional (3D) slopes in the selected area. It is convenient to extract slope gradients in ArcGIS based on the divided grid units. As shown in Fig. 7.11a, a cross-section is selected from Section 1 and named Cross-section 1. The slope ranges from units 57 to 71 are selected for analysis with a height of 80 m, as

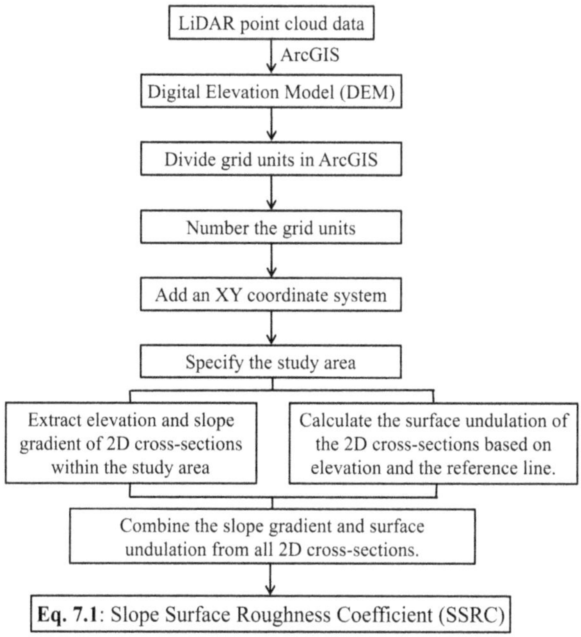

Fig. 7.10 The flowchart of the process for extracting SSRC data in ArcGIS

illustrated in Fig. 7.11b. A reference line, OR, with an angle of 45° is used for calculating surface undulation. The segment OH is divided into eight equal parts, and then surface undulation is calculated using Eqs. (7.2) and (7.3). This method can be applied to determine the surface undulation of any other two-dimensional cross-section. As shown in Fig. 7.12, ArcGIS can extract elevation and slope gradient for any arbitrary 2D cross-section, and the six arbitrarily selected 2D cross-sections have different elevations. For the current SSRC calculation, only regions of the three-dimensional slope with similar heights (variations within 10 m) can be analyzed, as this range has minimal impact on the safety factor of 3D slopes. However, for comparisons of safety factors between 2D cross-sections with significant height variations, future research will consider slope height as an additional factor affecting the FoS. For three-dimensional slopes with similar heights, a reference line can be selected for calculating the surface roughness, as shown in Fig. 7.11b.

$$\Delta H_{\mathrm{i}} = \mathrm{H}_{\mathrm{S}} - \mathrm{H}_{\mathrm{R}} \tag{7.2}$$

$$\Delta h_{\mathrm{i}} = \Delta H_{\mathrm{i}} \cdot \cos\alpha \tag{7.3}$$

Table 7.2 summarizes the elevations for grid units numbered 57 to 71. The slope gradient and surface undulation are as follows: an average gradient of 33° and a maximum gradient of 50°, an average surface undulation of 8.2 m, and the maximum undulation is 14.0 m. This demonstrates that the calculation for a single

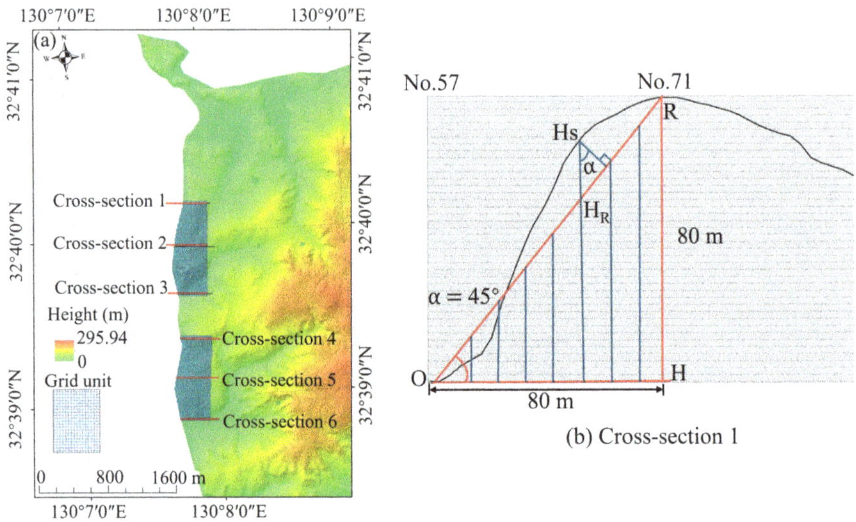

Fig. 7.11 (**a**) Six selected 2D cross sections, and (**b**) calculation diagram of Cross-section 1

two-dimensional (2D) cross-section in ArcGIS is relatively convenient compared to the analysis in Rhino. By combining the slope gradients and undulation data from a series of two-dimensional cross-sections, the slope surface undulation and gradient data for the three-dimensional (3D) slope can be obtained. The SSRC (Slope Surface Roughness Coefficient) can then be calculated based on the extracted slope and undulation data of the three-dimensional slope.

As shown in Fig. 7.13a, a 5 m × 5 m grid unit is applied to divide the specified areas for analysis. Section 1 which is 1060 m in length, and Section 2 which is 900 m in length. Each grid cell within the specified area is then numbered sequentially, as illustrated in Fig. 7.13b. This allows for the extraction of slope gradient and elevation information for any two-dimensional cross-section along the Ox direction. The extracted data is then imported into Excel for further processing. In the Ox direction, the combined elevations and gradients from units 1 to 28 yield the elevation and slope gradient data of a two-dimensional cross-section. In the Oy direction, the combined elevation and slope gradient of each two-dimensional cross-section result in the three-dimensional SSRC statistical data for the area. In Section 1, the grid units for the first cross-section, Cross-section 1.1 which ranges from unit 1 to unit 28. In contrast to Rhino where it is easy to set a unified reference plane to calculate the surface undulation of 3D slopes, it is challenging to set a unified reference plane in ArcGIS. Therefore, a method of setting a reference line for each individual two-dimensional cross-section was adopted in ArcGIS when performing statistical calculations on surface undulation. For this area, it is only necessary to separately record the elevation and slope gradient for numbered grid units 1 to 434, corresponding to 2D cross-sections 1 to 15 of Section 1 from top to bottom (Oy direction). By integrating the undulations and slope gradients calculated for these 2D

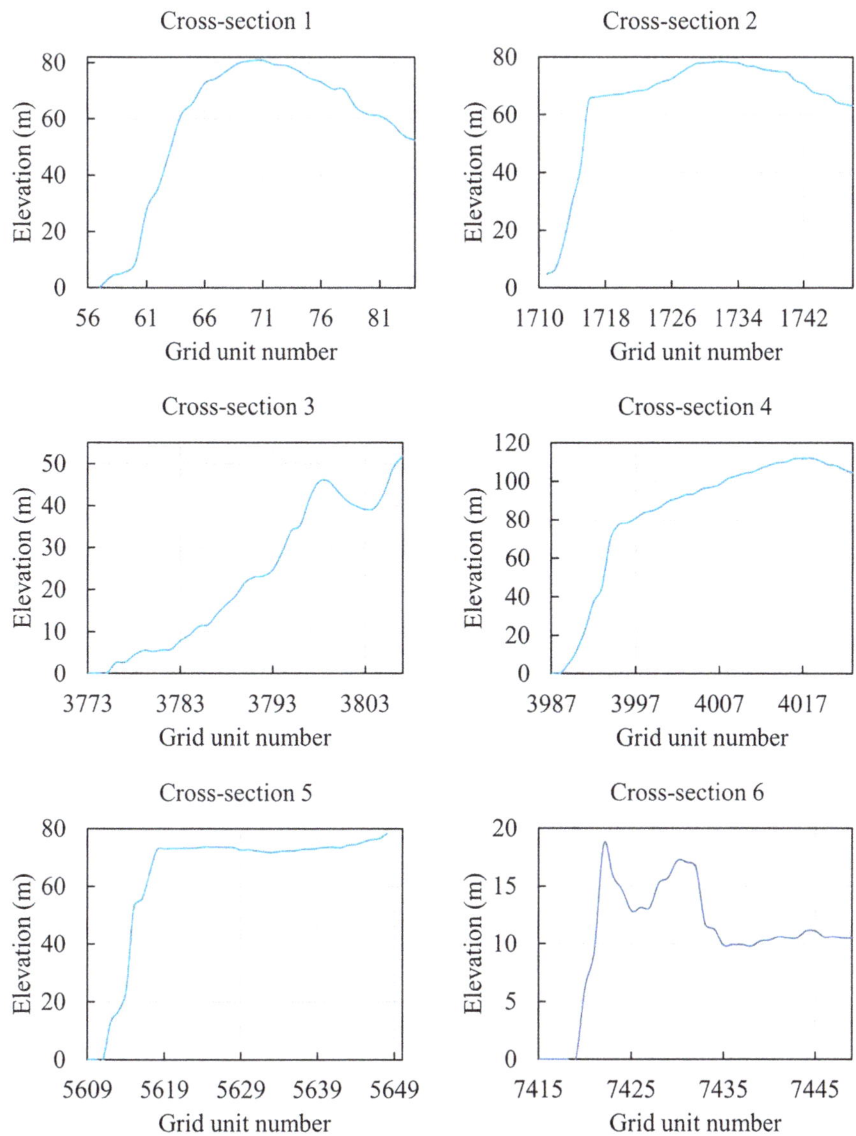

Fig. 7.12 Diagram of Cross-sections 1 to 6

Table 7.2 Elevation of cross-section 1

Grid No.	57	58	59	60	61	62	63	64	65	66	67	68	69	70	71
H_D	0	5	10	15	20	25	30	35	40	45	50	55	60	65	70
H_{Ri}	0	5.7	11.4	17.1	22.8	28.5	34.2	39.9	45.6	51.3	57	62.7	68.4	74.1	80
H_{Si}	0	3.9	5.3	8.5	27.8	35.3	48.2	61.2	65.7	72.7	74.4	77.3	79.9	80	80

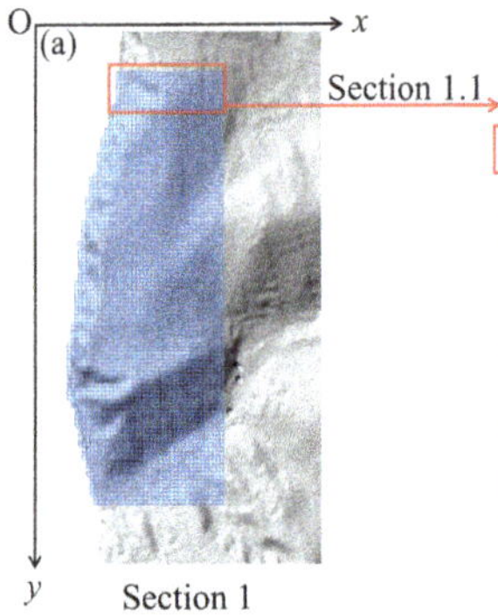

Cross-section 1

1	2	3	4	5	6	7	8	9	10	11	12	13	14	15
29	30	31	32	33	34	35	36	37	38	39	40	41	42	43
57	58	59	60	61	62	63	64	65	66	67	68	69	70	71
85	86	87	88	89	90	91	92	93	94	95	96	97	98	99
113	114	115	116	117	118	119	120	121	122	123	124	125	126	127
141	142	143	144	145	146	147	148	149	150	151	152	153	154	155
169	170	171	172	173	174	175	176	177	178	179	180	181	182	183
197	198	199	200	201	202	203	204	205	206	207	208	209	210	211
226	227	228	229	230	231	232	233	234	235	236	237	238	239	240
256	257	258	259	260	261	262	263	264	265	266	267	268	269	270
286	287	288	289	290	291	292	293	294	295	296	297	298	299	300
316	317	318	319	320	321	322	323	324	325	326	327	328	329	330
346	347	348	349	350	351	352	353	354	355	356	357	358	359	360
376	377	378	379	380	381	382	383	384	385	386	387	388	389	390

80 m

(b) Assign grid numbers

Fig. 7.13 (**a**) A 140 m × 200 m area in Section 1 (**b**) Numbering the area in ArcGIS.

cross-sections, the undulation and slope gradient of the entire 3D slope surface can be obtained. Table 7.3 summarizes the slope gradients and surface undulation for Cross-sections 1 to 15 within the specified study area.

Figure 7.14 shows the variation of slope gradients and surface undulations after integrating all these two-dimensional cross-sections. This method can be used to extract information of geometric features of three-dimensional slopes in any given area. However, it is important to emphasize that the undulation here is calculated using a reference line with a 45° gradient set for each cross-section, which differs from setting a unified reference plane in Rhino (as shown in Fig. 7.11b). In terms of accuracy, extracting slope gradients of 2D cross-sections in ArcGIS is more convenient and accurate compared to Rhino, while calculating surface undulation is easier in Rhino. It is obvious that the calculation method using ArcGIS becomes more effective and accurate as the area under analysis increases. However, the accuracy of surface undulation statistics using a reference line for a single two-dimensional cross-section still requires further investigation. Using custom development in ArcGIS to set a unified reference plane for calculating surface undulation will be explored in future research.

7.3.3 Analysis of 2D Cross-Sections

It is important to clarify that the curving slope surface examined in this study refers to curvature in the third dimension, which is perpendicular to the plane of interest. Detailed research of the curvature within the plane of interest is beyond the scope of this work. In general, the factor of safety (FoS) for convex slopes is smaller than that of common 3D slopes, while the FoS for concave slopes is larger than that of common 3D slopes. Therefore, for individual two-dimensional sections, only the convex features of the cross-section were included in the calculation of the SSRC. By

Table 7.3 Slope gradients and surface undulation of the target area

Cross-sections	1	2	3	4	5	6	7	8	9	10	11	12	13	14	15
Average gradient (°)	53	53	54	55	55	57	61	61	57	59	57	72	60	61	60
Maximum gradient (°)	22	25	27	25	23	23	21	21	22	22	19	33	19	17	17
Average undulation (m)	16.5	18.5	12.1	12.5	8.2	8.5	12.9	13.3	13.6	10.7	11.1	12.9	12.4	13.9	18.0
Maximum undulation (m)	9.7	11.2	5.7	5.5	5.2	5.7	7.6	8.4	7.0	4.8	5.0	8.2	6.9	7.7	10.3

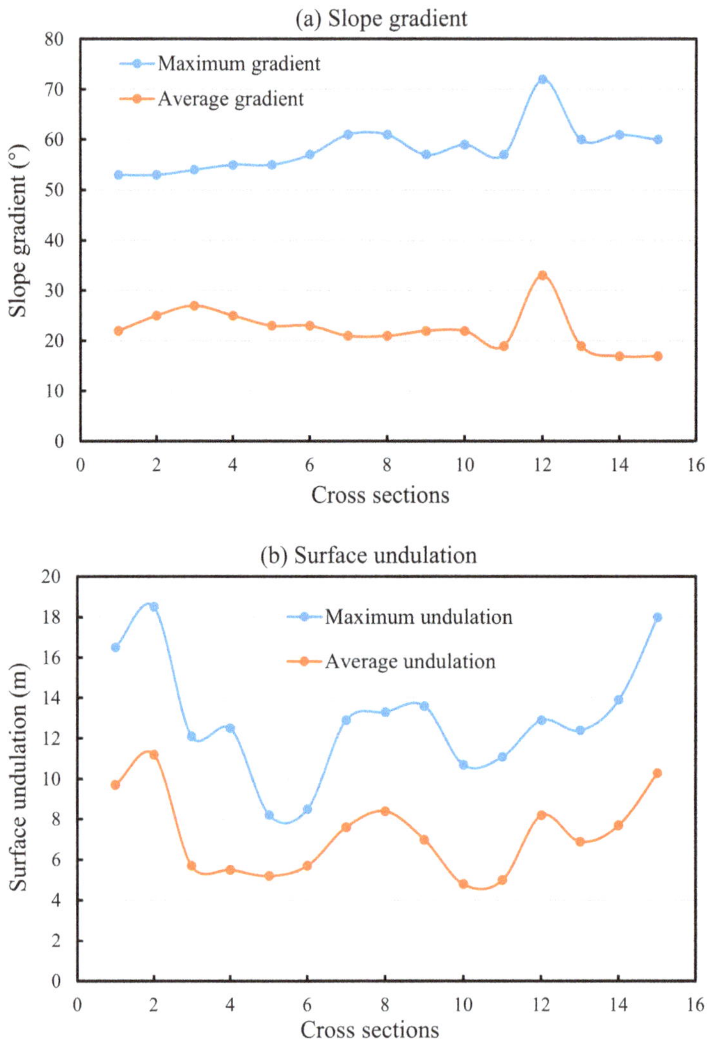

Fig. 7.14 (**a**) The slope gradient and (**b**) surface undulation of the area

integrating the surface undulation data extracted from all the 2D sections, the longitudinal undulation characteristics of the 3D slope can be determined. MATLAB programming was used to extract geometric information from all the two-dimensional cross-sections. For a single two-dimensional cross-section, it was also used to calculate all the surface undulations and local slope angles, as well as to compute the maximum and average values of the surface undulations and gradients. The tasks of automatically recording and calculating elevation and slope gradient of a 2D cross-section can be achieved based on the image recognition and processing functions of MATLAB.

As shown in Fig. 7.15, the horizontal distance OA is evenly divided into segments of 5 m (Δx_i). The elevation difference between two adjacent points where the segments intersect with the slope surface is denoted as $\Delta z_i = z_{i+1} - z_i$. The local angle θ_i for each segment is obtained by taking the inverse tangent of the ratio between the elevation difference Δz_i and the distance Δx_i, which is express as: $\theta_i = \tan^{-1}(\Delta z_i / \Delta x_i)$. The elevation difference between the slope surface (H_S) and the reference plane (H_R) is represented as $\Delta H_i = H_S - H_R$. The vertical distance between the slope surface and the reference line Δh_i, is calculated using the formula: $\Delta h_i = \Delta H_i \cdot \cos\alpha$, where α is the angle between the horizontal plane and the reference plane. Since the calculation of SSRC (Slope Surface Roughness Coefficient) only considers the data from the convex sections of the slope surface, the focus is on areas where the slope surface rises above the reference line. Therefore, the maximum undulation of a 2D section (Δh_{max}) is defined as the maximum value of vertical distance between the slope surface and the reference plane, as illustrated in Eq. (7.4). The average of these vertical distances is defined as the average slope surface undulation Δh_{avg}, as illustrated in Eq. (7.5). The integration of the surface undulations from all 2D sections reflects the longitudinal undulation characteristics of the 3D slope. The maximum slope gradient ($\Delta\theta_{max}$) of a 2D section is defined as the maximum value of all the local slope angle θ_i, as expressed in Eq. (7.6). The average angle of all these local angles is defined as the average slope gradient $\Delta\theta_{avg}$, as expressed in Eq. (7.7).

$$\Delta h_{\text{max}} = \max\left(\Delta h_i\right) \tag{7.4}$$

$$\Delta h_{\text{avg}} = \frac{1}{N-1}\sum_{i=1}^{N-1}\left|\Delta H_{i+1} - \Delta H_i\right| \tag{7.5}$$

$$\Delta \theta_{\text{max}} = \max\left(\theta_i\right) \tag{7.6}$$

$$\Delta \theta_{\text{avg}} = \frac{1}{N-1}\sum_{i=1}^{N-1}\left|\Delta \theta_{i+1} - \Delta \theta_i\right| \tag{7.7}$$

where N represents the total number of segments after dividing the slope into sections of 5 m each for a two-dimensional cross-section.

7.3.4 SSRC of Six Slope Models

Region 1, as shown in Fig. 7.8a, is further divided into three areas: Area 1, Area 2, and Area 3, each measuring 140 m in length. From these three areas, Slope models 1, 2, and 3 are extracted in the FLAC^{3D}, each measuring 140 m in length, 140 m in width, and 80 m in height, as illustrated in Fig. 7.16a–c. Similarly, Slope models 4, 5, and 6 are extracted from Region 2, as illustrated in Fig. 7.16d–f.

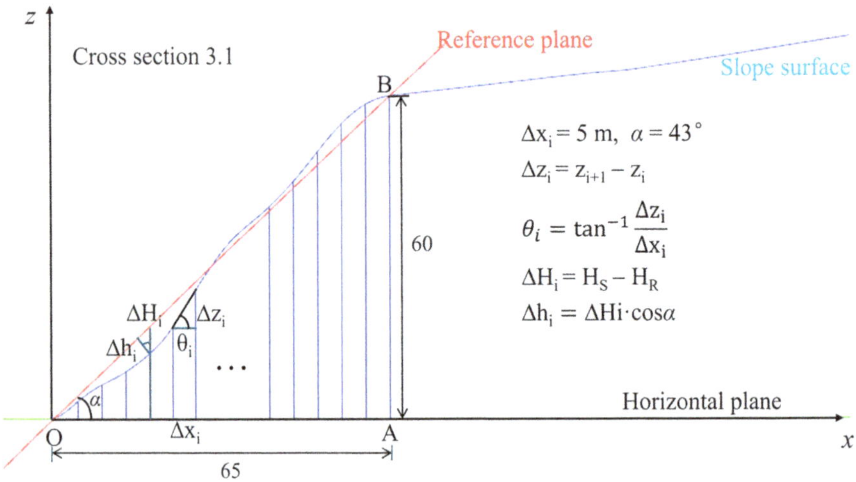

Fig. 7.15 A two-dimensional cross-section generated in Rhino (unit: m)

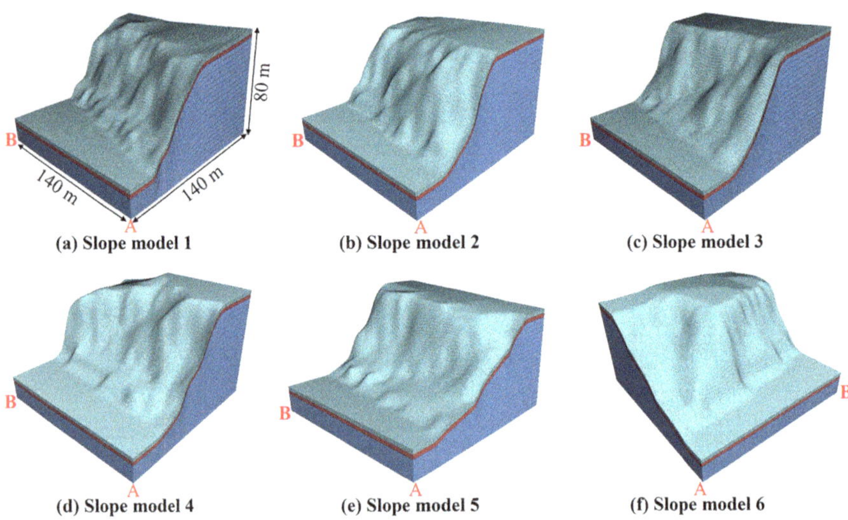

Fig. 7.16 Six slope models extracted from Regions 1 and 2

Figures 7.17 and 7.18 present the statistical data for the surface undulation and slope gradient of slope models 3, 4, 5, and 6. As shown in Fig. 7.17, it is evident that the maximum and average slope gradients for Model 3 are significantly lower than those for Model 4. However, the maximum and average surface undulation values for Model 3 are notably greater than those for Model 4. Figure 7.18 indicates that the maximum and average values of surface undulation and slope gradient for Model 5 are both lower than those for Model 6. The maximum slope gradient for Model 4 is significantly higher than the reference value of 52°, with the average value distributed around ±10° of the reference value (Fig. 7.17c). In contrast, the maximum slope gradient for Model 5 is mostly concentrated around the reference value of 52°, while the average slope gradient tends to be below this reference value (Fig. 7.18a). Table 7.4 summarizes the maximum and average value of surface undulation and slope gradient of these six slope models. In Table 7.4, the average value represents the average surface undulation and slope gradient of the three-dimensional slope, calculated as the mean of the average values from each two-dimensional cross-section. Meanwhile, the maximum value represents the maximum surface undulation and slope gradient of the three-dimensional slope, indicating the highest maximum value among all cross-sections. According to the previously established standard slope model (Model O4), significant changes in the safety factor only occur when the surface undulation of the slope reaches 10 m. The safety factor begins to change noticeably when the undulation frequency exceeds 2.

Table 7.4 presents the adjusted SSRC values for all the real three-dimensional slopes. For the purpose of standardization, the reference model is chosen as Model O4 for calculation of SSRC for real Slope models 1 to 6. Therefore, the reference undulation is set at 5 m, and the reference frequency is set at 2. Due to the

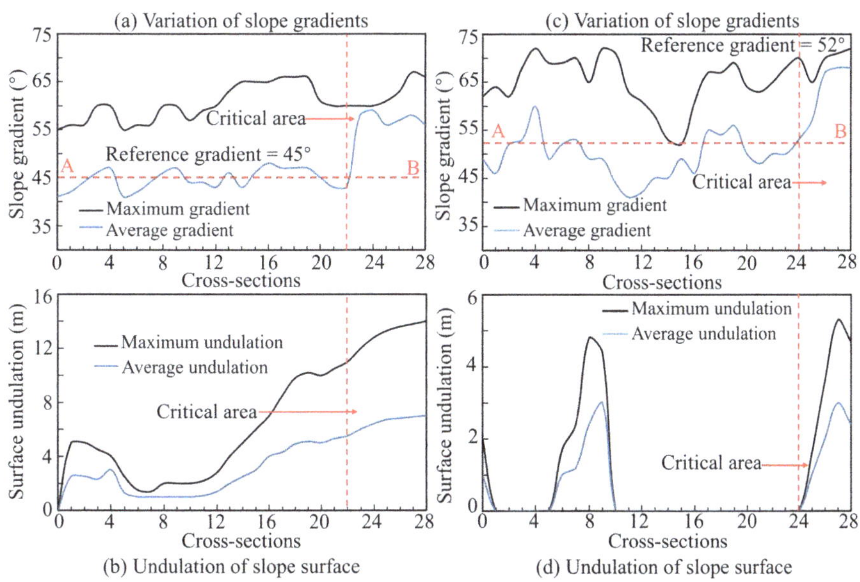

Fig. 7.17 The statistical data for the surface undulation (**a**) and slope gradient (**b**) of Model 3, and for the surface undulation (**c**) and slope gradient (**d**) of Model 4

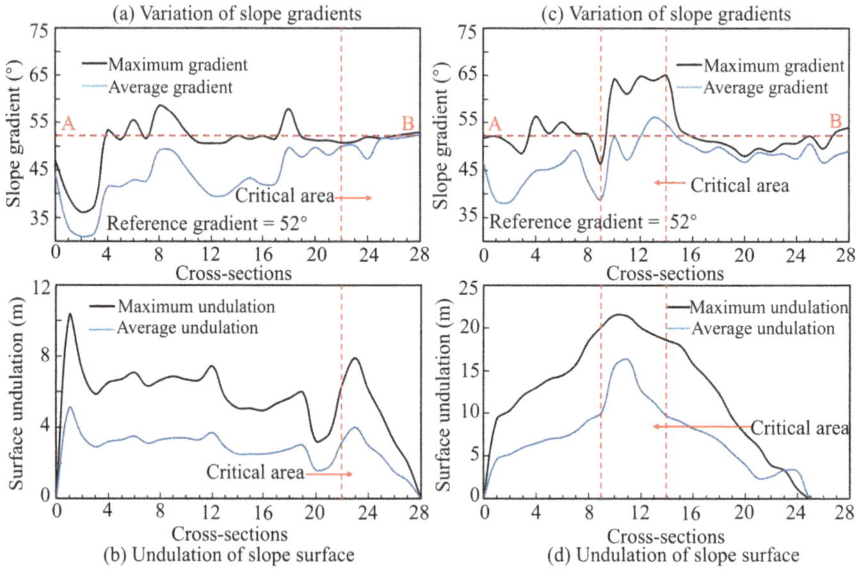

Fig. 7.18 The statistical data for the surface undulation (**a**) and slope gradient (**b**) of Model 5, and for the surface undulation (**c**) and slope gradient (**d**) of Model 6

differences in slope gradient, Slope models 1, 2, and 3 correspond to the reference Model Standard 1, with dimensions of 140 m in length, 140 m in width, and 80 m in height, and a slope gradient of 43°. Slope Models 4, 5, and 6 correspond to the reference Model Standard 2, with the same dimensions but a slope gradient of 52°. The calculated values of SSRC of Slope models 1 to 3 are evidently higher than the SSRC in Table 7.1. This is because the complex geometry of the actual slopes and the differences in the reference slopes, the gradient of Model O4 is 50°, while the gradients of Model standard 1 and standard 2 are 43° and 52°, respectively. Therefore, the SSRC of Slope models 1 to 6 cannot be directly used to calculate the predicted values of FoS for the real slope models. Here, the SSRC values of Slope models 1 to 3 were reduced by the factor of $\lambda_1 = 1.6$ according to the difference in slope gradient and the influence of error factors, while those of Slope models 4 to 6 were adjusted by the factor of $\lambda_2 = 1.1$. Due to variations in slope gradients and undulation degrees in the standard slope models, despite the SSRC for Model O2, Standard1, and Standard2 all being 2.46, adjustments are made to reflect these differences when evaluating 3D slope stability. It is worth emphasizing that once a sufficient amount of data is accumulated, the values of these adjustment factors will become more precise, and the predicted safety factor values will improve as the accuracy of SSRC increases.

Table 7.4 Results of SSRC for the slope models 1 to 6

Slope model	Average gradient $\Delta\theta_{avg}$ (°)	Maximum gradient $\Delta\theta_{max}$ (°)	Reference gradient $\Delta\theta_{avg}^{Ref}$ / $\Delta\theta_{max}^{Ref}$ (°)	Average undulation Δh_{avg} (m)	Maximum undulation Δh_{max} (m)	Reference undulation Δh_{avg}^{Ref} / Δh_{max}^{Ref} (m)	f_q (frequency of undulation≥10 m)	f_q^{Ref}	SSRC
1	49	69	43	3.9	10.6	5	5	1	2.50
2	53	70		8.0	21.2		6		2.77
3	47	67		3.5	14.6		11		2.61
Standar1	43	43		5	5		3		2.46
4	52	72	52	2.1	5.3	5	3		2.90
5	45	59		3.7	10.3		8		2.66
6	48	65		6.3	21.5		9		3.10
Standar2	52	52		5	5		1		2.46

7.4 Results

The complexity of slope geometry highlights the importance of utilizing robust analytical methods to account for the geometry characteristics of irregular slopes when evaluating their stability under rainfall conditions. Assuming that the slope heights, geotechnical parameters, and groundwater levels are consistent across all cases, the safety factor (FoS) of the three-dimensional slope can be evaluated based on the linear relationship between SSRC and FoS. The safety factor (FoS) can be evaluated for pre-rainfall conditions, as well as 24 h and 48 h after rainfall. Table 7.5 presents the predicted safety factors (FoS) for the three-dimensional slopes under heavy rainfall, specifically with a rainfall intensity of 20 mm/h.

In addition to simulating unsaturated flow in soil and the water-induced weakening effects of rocks in FLAC^{3D}, the fluid-solid coupling of unsaturated soils during the rainfall infiltration process was also implemented in FLAC^{3D}. Three-dimensional slope stability analysis of elastic-perfectly plastic soils and rocks with a Mohr-Coulomb failure criterion assuming zero dilation was performed, and no tensile strength was considered in the present research. The matric suction was assumed to be constant at a specific matric suction value above the maximum matric suction (−50 kPa) based on the study of Leong and Rahardjo (2012). Taking Slope model 1 as an example, the model is divided into a grid with a resolution of 1 m, resulting in a total of 509,600 zones and 536,787 grid points. Figure 7.19 illustrates the unsaturated seepage simulation in FLAC^{3D} based on the theoretical formulas presented in Chap. 3. The porosity of the moderately weathered layer was set at 0.14. and a conservative reduction factor λ of 0.8 were applied to weathered rocks.

As shown in Figs. 7.20 and 7.21, the areas of maximum shear strain (potential sliding zones) after 24 h of rainfall are predominantly located in the most convex and steepest parts of the slope. The scale of the critical areas in Figs. 7.20b and 7.21b closely corresponds to the potential sliding zones depicted in Figs. 7.20d and 7.21d, respectively. This illustrates that the potential sliding surfaces of a slope can be effectively analyzed under rainfall conditions by focusing on its geometric characteristics. This method is based on the assumption that other important factors, such as slope height, thickness of weathered layers, and groundwater level are remain consistent. Consequently, this method provides a solid foundation for

Table 7.5 Predicted FoS for the Slope models 1 to 6

	FoS		
	Rainfall duration (h)		
Slope model	0	24	40
1	2.19	1.60	1.38
2	1.98	1.47	1.29
3	2.11	1.55	1.35
4	1.90	1.41	1.25
5	2.07	1.52	1.33
6	1.73	1.32	1.18

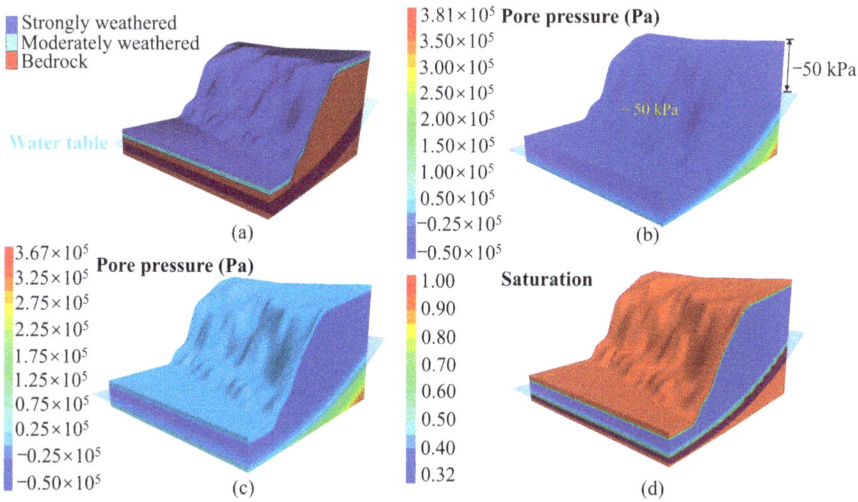

Fig. 7.19 (**a**) Slope model 1 with groundwater table in FLAC^{3D}. (**b**) The distribution of pore pressure before rainfall within the slope. (**c**) The distribution of pore pressure after 24 h of rainfall. (**d**) The distribution of saturation after 24 h of rainfall

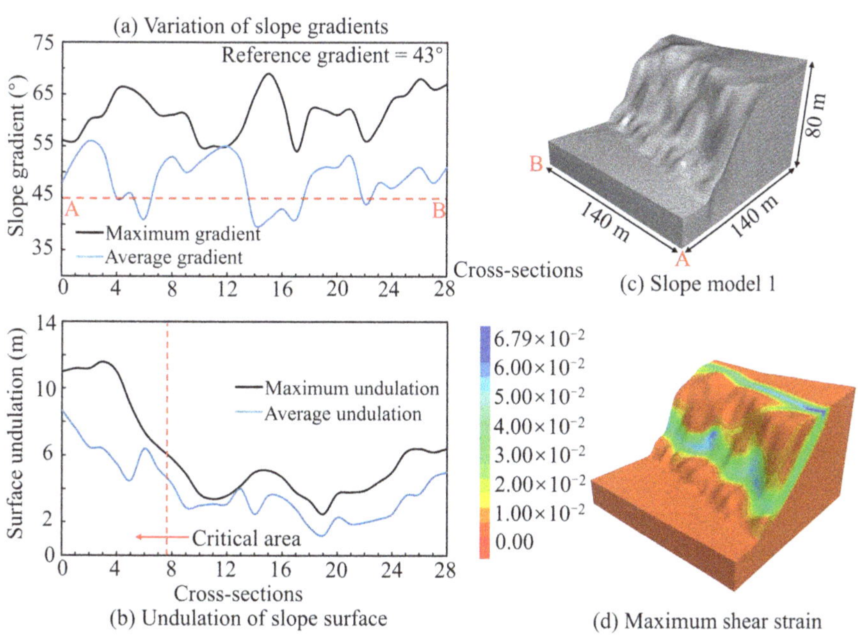

Fig. 7.20 The statistical data of surface undulation (**a**) and slope gradient (**b**) for Slope model 1, including its dimensions (**c**), and (**d**) the maximum shear strain after 24 h of rainfall

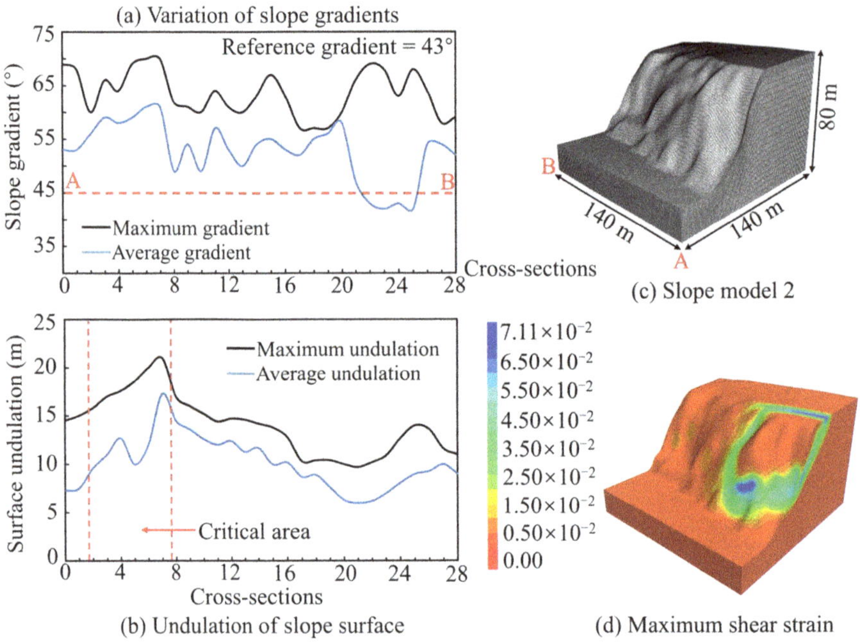

Fig. 7.21 The statistical data of surface undulation (**a**) and slope gradient (**b**) for Slope model 2, including its dimensions (**c**), and (**d**) the maximum shear strain after 24 h of rainfall

evaluating slope safety factors of 3D slopes based on the established SSRC. Subsequent research will be conducted on the impact of other factors on the 3D slope safety factor in the future.

There is a strong correlation between these zones and the critical areas identified in Figs. 7.20 and 7.21, highlighting the most critical regions are the convex and steep sections of the 3D slopes under rainfall conditions. Notably, Slope model 6 exhibits a pronounced undulation with a steep gradient in its central section. Figure 7.22a shows the predicted safety factor values (P) for Slope models 1 to 6 before rainfall, 24 h, and 48 h after rainfall. Figure 7.22b presents the corresponding numerical simulation results (R). A clear linear relationship between SSRC and FoS is evident from the comparison. Slope model 1 has the smallest SSRC value, indicating that the three-dimensional slope's geometric features are less pronounced, with a gentler slope gradient and minimal surface undulation. As a result, the safety factor of Slope model 1 is highest among all the six slope models. In contrast, Slope model 6 has the largest SSRC value, reflecting more pronounced geometric features with a steeper slope gradient and significant surface undulation. This leads to the smallest safety factor of Slope model 6 among all the six slope models. As illustrated in Fig. 7.18c, d, the critical area is approximately located between the 9th and 15th two-dimensional cross-sections, and the maximum shear strain region depicted in Fig. 7.23d corresponds well with this range. Table 7.6 summarizes the results of safety factor (FoS) calculated in $FLAC^{3D}$ for various rainfall durations. Therefore, it

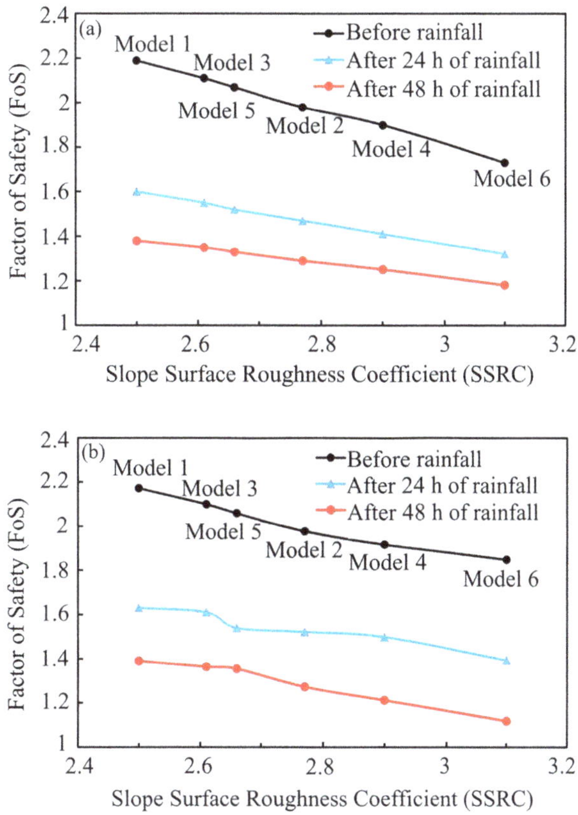

Fig. 7.22 The relationship between SSRC and FoS: (**a**) predicted values and (**b**) real values

is clear that we can use the slope gradient and surface undulation of a three-dimensional slope which is obtained from ArcGIS or Rhino to predict the safety factor of the three-dimensional slope.

To evaluate the accuracy of the model, two indices which are mean error (ME) and root mean square error (RMSE) were calculated from real values and predicted values among various slope models. ME is defined by Eq. (7.8). It measures the bias of the predicted value and should be close to zero for unbiased methods. RMSE is determined by Eq. (7.9). It reflects the absolute fit of the model to the data, indicating how closely the actual data aligns with the predicted values from the model. The calculated values for the ME and RMSE are low and acceptable from a statistical perspective (ME = 0.019, RMSE = 0.053).

$$ME = \frac{\sum_i \left[(z_i - s_i) \right]}{n} \tag{7.8}$$

$$RMSE = \left\{ \frac{\sum_i \left[(z_i - s_i) \right]^2}{n} \right\}^{1/2} \tag{7.9}$$

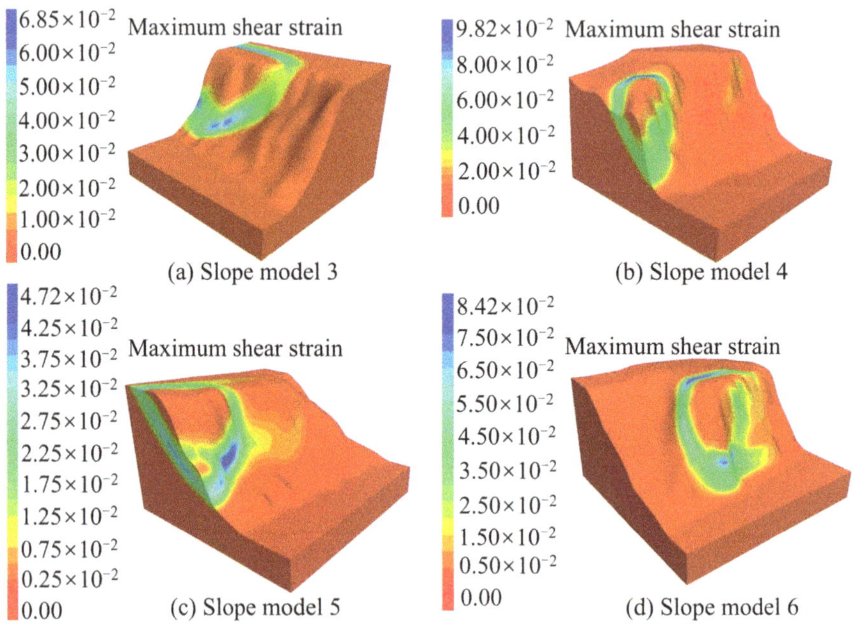

Fig. 7.23 The maximum shear strain of Slope models 3 to 6 after 24 h of rainfall

Table 7.6 The real FoS of Slope models 1 to 6

Slope model	FoS		
	Rainfall duration (h)		
	0	24	48
1	2.173	1.63	1.364
2	1.977	1.539	1.273
3	2.06	1.61	1.355
4	1.917	1.498	1.212
5	2.10	1.521	1.39
6	1.849	1.392	1.118

where z_i and s_i are the observed value and predicted value of ith validation point. n is the number of validation points.

The error analysis indicates that the predicted values closely align with the actual values, demonstrating the accuracy of the linear relationship established between SSRC and FoS. Thus, employing SSRC to assess changes in safety factors for three dimensional slopes under rainfall conditions is considered reliable. In this way, it is only necessary to calculate the FoS of a standard 3D slope under rainfall conditions. By conducting a comparative analysis of the SSRC between real and standard 3D slopes, the FoS of the actual 3D slope can be determined. This significantly alleviates the computational burden on numerical simulation in FLAC^{3D}, as the detailed transient seepage analysis of rainfall infiltration is performed exclusively for the standard three-dimensional models. This approach allows for an accurate and efficient evaluation of slope stability within a specific area under short-term heavy rainfall. After accumulating a large amount of computational data, machine learning models can be used to optimize the weights in the calculation of SSRC, taking into account multiple influencing factors affecting the FoS of the three-dimensional slope under rainfall conditions.

7.5 Discussions and Conclusions

This study is based on a detailed analysis of unsaturated infiltration in 3D slopes under rainfall conditions in $FLAC^{3D}$. It aims to accurately capture the variation in safety factors (FoS) of standard three-dimensional slopes during rainfall infiltration. An evaluation metric (SSRC) is proposed to account for the impact of surface undulation and slope gradient on the stability of 3D slopes. By utilizing GIS and Rhino to statistically analyze the geometric characteristics of actual 3D slopes and calculating their SSRC values, the linear relationship between SSRC and FoS can be used to assess the FoS of actual slopes based on the standard models. Error analysis between the predicted and actual results highlights the potential of applying this method for assessing the susceptibility of regional rainfall-induced landslides.

Although GIS has unique advantages in assessing the stability of rainfall-induced landslides on a regional scale, it typically uses one-dimensional GA model or Richards' equations for rainfall infiltration analysis in the slope. Such simplifications make it difficult to capture the detailed behavior of unsaturated seepage within soils and the further infiltration of rainwater into rock layer. In particular, the real-time changes in matric suction of soils, and their influence on shear strength parameters such as cohesion c and elastic modulus E during the rainfall infiltration, are often not addressed. GIS-based methods prioritize the efficient dealing with regional landslide hazard analysis, but at the cost of deeply analyzing the influence of unsaturated seepage on the stability of 3D slopes. Considering the changes in matric suction in unsaturated soils and the water-induced weakening effects of rocks, this study focuses on detailed calculations in $FLAC^{3D}$ to evaluate how rainfall infiltration affects the factor of safety (FoS) of 3D slopes. Given the computational constraints, the slope model selected for analysis is designed with dimensions of 140 m in length, 140 m in width, and 80 m in height. Based on the geometric characteristics of the three-dimensional slope (surface undulation and slope gradient) in a designated area, a standard 3D model is used to construct the SSRC (Slope Surface Roughness Coefficient), an index for evaluating the geometric characteristics of three-dimensional slopes. The relationship between SSRC and FoS is then utilized to evaluate the actual safety factor of the 3D slope.

Due to the current limited amount of data, the scope of evaluation is constrained to areas within a range of 420 m in length and 200 m in width. However, the prediction range can be expanded when enough amount of data is accumulated in the future. The current research establishes a solid foundation for future studies employing machine learning techniques for large-scale assessment of rainfall-induced landslide susceptibility. Trustworthy training and prediction rely on the accurate data of FoS for 3D slopes under heavy rainfall, which is the primary objective of this study. In addition to the SSRC proposed in this research, slope angle and surface undulation are just two significant factors influencing slope stability under rainfall conditions. Other important factors, such as cohesion c, internal friction angle φ, slope height and groundwater table (GWT), can also be incorporated into future machine learning training once a substantial amount of data is accumulated. The differences in the choice of reference line and their impact on the SSRC results will be further explored in subsequent research. The conclusions drawn from this study are as follows: By utilizing GIS to divide slope units based on ridge and valley lines, it is possible to roughly estimate the degree of surface undulation of 3D slopes over a given length (420 m). Combined with slope gradient and height extracted in GIS, this method can approximately identify regions that are prone to landslides. This is consistent with the results in Table 7.4, which show the safety factors of Slope models 1, 2, and 3, extracted from the relatively smooth surface of Region 1, are higher than those of Slope models 4, 5, and 6, which were extracted from the more undulating surface of Region 2.

The safety factor of three-dimensional slopes is strongly influenced by the variation in slope gradient along the longitudinal direction and the prominence of convex areas relative to the reference surface. The steeper the slope and the more pronounced the convex features, the more likely it is for potential sliding zones to develop in those areas. A clear linear relationship exists between SSRC and FoS, based on factors such as the maximum and average slope gradient, surface undulation, and the frequency of convex areas exceedingly more than 10 m above the reference plane in the longitudinal direction. A larger value of SSRC indicates a more geometrically complex slope and corresponds to a lower safety factor, while a smaller SSRC suggests simpler geometry and smoother surfaces, correlating with a higher safety factor. This relationship validates the reliability of using SSRC to assess the stability of 3D slopes under heavy rainfall. Once sufficient computational data is accumulated, research will focus on the assessment of large-scale slope stability under rainfall conditions based on the geometric features of the slope. The objective is to further develop and refine assessment techniques by incorporating advanced GIS methodologies and machine learning models to predict regional slope stability under various rainfall conditions. This includes gaining proficiency in GIS software and tools specific to slope stability analysis, understanding the application of GIS in the context of large-scale geohazard assessments, and developing machine learning models that predict slope stability based on geometric and rainfall data. The models will be validated and tested using historical data and real-world scenarios.

References

Barton N, Choubey V (1977) The shear strength of rock joints in theory and practice. Rock Mech 10:1–54. https://doi.org/10.1007/BF01261801
Baum RL, Godt JW, Savage WZ (2010) Estimating the timing and location of shallow rainfall-induced landslides using a model for transient, unsaturated infiltration. J Geophys Res 115:2009JF001321. https://doi.org/10.1029/2009JF001321
Bishop AW (1955) The use of the slip circle in the stability analysis of slopes. Géotechnique 5:7–17. https://doi.org/10.1680/geot.1955.5.1.7
Cai F, Ugai K (2004) Numerical analysis of rainfall effects on slope stability. Int J Geomech 4:69–78. https://doi.org/10.1061/(ASCE)1532-3641(2004)4:2(69)
Cai X, Zhou Z, Zang H, Song Z (2020) Water saturation effects on dynamic behavior and micro-structure damage of sandstone: phenomena and mechanisms. Eng Geol 276:105760. https://doi.org/10.1016/j.enggeo.2020.105760
Gu T, Wang J, Fu X, Liu Y (2015) GIS and limit equilibrium in the assessment of regional slope stability and mapping of landslide susceptibility. Bull Eng Geol Environ 74:1105–1115. https://doi.org/10.1007/s10064-014-0689-2
Gutiérrez-Martín A (2020) A GIS-physically-based emergency methodology for predicting rainfall-induced shallow landslide zonation. Geomorphology 359:107121. https://doi.org/10.1016/j.geomorph.2020.107121
Heber Green W, Ampt GA (1911) Studies on soil physics. J Agric Sci 4:1–24. https://doi.org/10.1017/S0021859600001441
Ho J-Y, Lee KT, Chang T-C et al (2012) Influences of spatial distribution of soil thickness on shallow landslide prediction. Eng Geol 124:38–46. https://doi.org/10.1016/j.enggeo.2011.09.013
Hovland HJ (1977) Three-dimensional slope stability analysis method. J Geotech Eng Div 103:971–986. https://doi.org/10.1061/AJGEB6.0000493
Hoyos LR, Suescún-Florez EA, Puppala AJ (2015) Stiffness of intermediate unsaturated soil from simultaneous suction-controlled resonant column and bender element testing. Eng Geol 188:10–28. https://doi.org/10.1016/j.enggeo.2015.01.014
Iverson RM (2000) Landslide triggering by rain infiltration. Water Resour Res 36:1897–1910. https://doi.org/10.1029/2000WR900090
Iyare UC, Blake OO, Frash LP et al (2023) Water-weakening effects on the failure behavior of mudstones. Rock Mech Rock Eng 56:9171–9185. https://doi.org/10.1007/s00603-023-03544-4
Jia N, Mitani Y, Xie M, Djamaluddin I (2012) Shallow landslide hazard assessment using a three-dimensional deterministic model in a mountainous area. Comput Geotech 45:1–10. https://doi.org/10.1016/j.compgeo.2012.04.007
Jia N, Mitani Y, Xie M et al (2015) GIS deterministic model-based 3D large-scale artificial slope stability analysis along a highway using a new slope unit division method. Nat Hazards 76:873–890. https://doi.org/10.1007/s11069-014-1524-6
Lee Y-F, Chi Y-Y (2011) Rainfall-induced landslide risk at Lushan, Taiwan. Eng Geol 123:113–121. https://doi.org/10.1016/j.enggeo.2011.03.006
Leong EC, Rahardjo H (2012) Two and three-dimensional slope stability reanalyses of Bukit Batok slope. Comput Geotech 42:81–88. https://doi.org/10.1016/j.compgeo.2012.01.001
Li D, Wong LNY, Liu G, Zhang X (2012) Influence of water content and anisotropy on the strength and deformability of low porosity meta-sedimentary rocks under triaxial compression. Eng Geol 126:46–66. https://doi.org/10.1016/j.enggeo.2011.12.009
Li X, Jiang Y, Sugimoto S (2024) Research on the influence of spatial dimensions on the stability of large-scale slopes under heavy rainfall. GeoHazards 5:732–754. https://doi.org/10.3390/geohazards5030037
Ng CWW, Shi Q (1998) A numerical investigation of the stability of unsaturated soil slopes subjected to transient seepage. Comput Geotech 22:1–28. https://doi.org/10.1016/S0266-352X(97)00036-0

Nian T-K, Huang R-Q, Wan S-S, Chen G-Q (2012) Three-dimensional strength-reduction finite element analysis of slopes: geometric effects. Can Geotech J 49:574–588. https://doi.org/10.1139/t2012-014

Pan Y, Wu G, Zhao Z, He L (2020) Analysis of rock slope stability under rainfall conditions considering the water-induced weakening of rock. Comput Geotech 128:103806. https://doi.org/10.1016/j.compgeo.2020.103806

Palazzolo N, Peres D, Bordoni M, et al (2021) Improving Spatial Landslide Prediction with 3D Slope Stability Analysis and Genetic Algorithm Optimization: Application to the Oltrepò Pavese. Water 13:801. https://doi.org/10.3390/w13060801

Park HJ, Lee JH, Woo I (2013) Assessment of rainfall-induced shallow landslide susceptibility using a GIS-based probabilistic approach. Eng Geol 161:1–15. https://doi.org/10.1016/j.enggeo.2013.04.011

Segoni S, Rossi G, Catani F (2012) Improving basin scale shallow landslide modelling using reliable soil thickness maps. Nat Hazards 61:85–101. https://doi.org/10.1007/s11069-011-9770-3

Tozato K, Dolojan NLJ, Touge Y et al (2022) Limit equilibrium method-based 3D slope stability analysis for wide area considering influence of rainfall. Eng Geol 308:106808. https://doi.org/10.1016/j.enggeo.2022.106808

Tran TV, Alvioli M, Lee G, An HU (2017) Three-dimensional, time-dependent modeling of rainfall-induced landslides over a digital landscape: a case study. Landslides 15:1071–1084. https://doi.org/10.1007/s10346-017-0931-7

Tran TV, Alvioli M, Hoang VH (2022) Description of a complex, rainfall-induced landslide within a multi-stage three-dimensional model. Nat Hazards 110:1953–1968. https://doi.org/10.1007/s11069-021-05020-0

Van Den Bout B, Lombardo L, Chiyang M et al (2021) Physically-based catchment-scale prediction of slope failure volume and geometry. Eng Geol 284:105942. https://doi.org/10.1016/j.enggeo.2020.105942

Van Westen CJ, Van Asch TWJ, Soeters R (2006) Landslide hazard and risk zonation—why is it still so difficult? Bull Eng Geol Environ 65:167–184. https://doi.org/10.1007/s10064-005-0023-0

Vásárhelyi B, Ván P (2006) Influence of water content on the strength of rock. Eng Geol 84:70–74. https://doi.org/10.1016/j.enggeo.2005.11.011

Wei WB, Cheng YM, Li L (2009) Three-dimensional slope failure analysis by the strength reduction and limit equilibrium methods. Comput Geotech 36:70–80. https://doi.org/10.1016/j.compgeo.2008.03.003

Xie M, Esaki T, Cai M (2004) A GIS-based method for locating the critical 3D slip surface in a slope. Comput Geotech 31:267–277. https://doi.org/10.1016/j.compgeo.2004.03.003

Xie M, Tetsuro E, Qiu C, Jia L (2007) Spatial three-dimensional landslide susceptibility mapping tool and its applications. Earth Sci Front 14:73–84. https://doi.org/10.1016/S1872-5791(08)60004-4

Xu Y, Li J, Fan H et al (2017) Stability analysis of clastic rock slope with mudstone interlayer under rainfall infiltration. Geotech Geol Eng 35:1871–1883. https://doi.org/10.1007/s10706-017-0215-y

Zhang Y, Chen G, Wang B, Li L (2013a) An analytical method to evaluate the effect of a turning corner on 3D slope stability. Comput Geotech 53:40–45. https://doi.org/10.1016/j.compgeo.2013.05.002

Zhang Y, Chen G, Zheng L et al (2013b) Effects of geometries on three-dimensional slope stability. Can Geotech J 50:233–249. https://doi.org/10.1139/cgj-2012-0279

Zhang LL, Fredlund MD, Fredlund DG et al (2015) The influence of the unsaturated soil zone on 2-D and 3-D slope stability analyses. Eng Geol 193:374–383. https://doi.org/10.1016/j.enggeo.2015.05.011

Zheng H (2012) A three-dimensional rigorous method for stability analysis of landslides. Eng Geol 145–146:30–40. https://doi.org/10.1016/j.enggeo.2012.06.010

Zhou Y, Qi SC, Fan G et al (2020) Topographic effects on three-dimensional slope stability for fluctuating water conditions using numerical analysis. Water 12:615. https://doi.org/10.3390/w12020615

Zienkiewicz OC, Humpheson C, Lewis RW (1975) Associated and non-associated visco-plasticity and plasticity in soil mechanics. Géotechnique 25:671–689. https://doi.org/10.1680/geot.1975.25.4.671

Index

Y. Jiang, X. Li, *Stability Evaluation of Slope Under Heavy Rainfall Based on Geometric Features*, Springer Natural Hazards,
https://doi.org/10.1007/978-981-95-8656-1